Chemistry 111L/112L
Laboratory Experiments

**UNIVERSITY
OF
SOUTH CAROLINA**

Department of Chemistry and Biochemistry

D. L. FREEMAN

D. REGER

A. TAYLOR-PERRY

Copyright © 2018 by Freeman, Reger, Taylor-Perry
Copyright © 2018 Illustrations by QDE Press Inc.

All rights reserved.

Permission in writing must be obtained from the publisher before any part of this work may be reproduced or transmitted in any form or by any means, electronic or mechanical, including photocopying and recording, or by any information storage or retrieval system.

Printed in the United States of America.

ISBN: 978-1-938535-18-5

QDE Press Inc.
8828 Autumnbrooke Way
Montgomery, Al 36117
www.qdepress.com

Chemistry 111L/112L Laboratory Experiments

Page

Chemistry 111L

Exp. 1 Safety and Laboratory Techniques	1
Exp. 2 Physical Properties of Substances	11
Exp. 3 Percent of Copper in Copper(II) Sulfate Pentahydrate	19
Exp. 4 Limiting Reactant and Percent Yield	27
Exp. 5 Acid-Base Titration: Determining the Conc. of a NaOH Solution	35
Exp. 6 Heats of Formation: Hess's Law	43
Exp. 7 The Ideal Gas Law: Pressure-Temperature Relationships	53
Exp. 8 Molar Mass by Vapor Density	65
Exp. 9 Experimental Determination of R, The Ideal Gas Constant	71
Exp. 10 Paper Chromatography	81
Exp. 11 Determination of Waters of Hydration	89
Exp. 12 Shapes of Molecules	95

Chemistry 112L

Safety Policies	105
Exp. 13 Qualitative Inorganic Analysis	109
Exp. 14 Molar Mass Determination by Freezing Point Depression	139
Exp. 15 Diprotic Acid: Identifying an Unknown by Titration	149
Exp. 16 The Acid Ionization Constant (K_a) of Acetic Acid	159
Exp. 17 Identification of an Unknown Acid by Titration	167
Exp. 18 Identification of Metals by Measuring Potentials of Micro-Voltaic Cells	179
Exp. 19 The Solvay Process - Preparation of Sodium Bicarbonate	189
Exp. 20 Determining the Rate Law for the Reaction of Crystal Violet with (OH^-)	195

Experiment 1
Safety and Laboratory Techniques

Objective

To become familiar with general safety rules, working safely with chemicals and hazard identifications in the laboratory. You will also learn to use some of the equipment found in the laboratory.

General Safety Rules
The following rules must be observed while working in the laboratory.

1. **Appropriate safety glasses must be worn at all times** – the use of contact lenses is also discouraged, however, if you do plan to wear contact lenses in the laboratory you must inform your instructor and wear safety **goggles**.

2. **Proper dress is required** – Bare feet, sandals, or opened-toed shoes are not allowed in the laboratory. It is best not to wear expensive clothing as stains and holes can result from misplaced chemicals.

3. **Food and drinks of any kind are not allowed in the Laboratory** – Keep all objects, such as glassware or plastic tubing, out of your mouth while in the lab.

4. **Behaving appropriately** – When working in the laboratory, you must be aware of others around you. Students will be carrying chemicals to and from their workstation, so be careful when walking through the lab. **Always act responsible in the lab.**

5. **Cleaning up** – It is important that you clean your work station upon completing your laboratory exercise. Make sure the gas is turned off to the Bunsen burners, remove any paper towels to the trash bin and clean up any spilled chemicals.

Experiment 1   Safety and Laboratory Techniques

6. Learn the location and operation of the safety equipment – Your laboratory is equipped with safety showers, eyewashes and fire extinguishers. You should become familiar with the location of each of these items as well as the location of all exits. If the fire alarm goes off while you are in the laboratory turn off all open flames, and follow the instructions of your laboratory instructor.

7. Safety video – Before the first experiment begins all students will watch a safety video outlining the topics discussed above as well as a description of proper handling of chemicals and chemical spills. After watching the film each student will fill out and sign the safety sheet found at the end of this chapter. All students must also complete the contact lens form.

Laboratory Techniques
In this section you will learn the proper technique for using some common laboratory equipment.

At your station you will find a graduated cylinder, buret, Bunsen burner and a 50-mL beaker. Your instructor will demonstrate the proper technique for reading the graduated cylinder and the buret as well as how to properly use a Bunsen burner and the analytical balances. After your instructor demonstrates these techniques you must demonstrate your ability to use this equipment to your instructor by completing the following exercise.

Competency Exercise
Light your Bunsen burner. Adjust the amount of air that is mixed with the gas so that a small bright blue cone is in the center of the flame and no "yellow" flame is visible. Show the flame to your instructor.

Fill the graduated cylinder about half full of tap water. Record the volume of water (to one decimal place) in the Data section and ask your instructor to verify your answer.

Fill the buret with water to near the zero mark. Drain a little of the water by rotating the stopcock. Make sure that no air bubbles are in the tip and the water level is below the zero mark. Read the volume off the buret and record your answer in the data section. In this measurement, you can record two decimal places because the buret is more exactly calibrated than the graduated cylinder. Using the analytical balance, record the mass of the empty 50-mL beaker. Using the water in the buret, fill the 50-mL beaker about quarter full. Read the new volume off the buret and record the value, again to two decimal places. Using the balance, determine the mass of the 50-mL beaker and added water and record the results. Calculate the volume and mass of water placed into the 50-mL beaker and record your answers on the data sheet. Be careful to record the values with the correct number of significant figures, based on the significant figures rules for subtraction. Given that the **density** of a substance is its mass divided by its volume, calculate the density of the water. Record your answer on the data sheet with the correct number of significant figures based on the rules for division.

Experiment 1 Safety and Laboratory Techniques

Worksheet 1

Name _____

Date _____ Lab Instructor/Section _____

Safety Quest.	____/20
Data	____/20
Drawing	____/20
Safety/Part.	____/40
Total	____/100

Data sheet

	Data	Instructor Initials
1. Bunsen Burner lighting	Not applicable	
2. Volume of Water Graduated cylinder	mL	
3. Initial reading of buret	mL	
4. Reading of the buret after filling the beaker	mL	
4. Mass of the empty 50-mL beaker	g	
5. Mass of the 50-mL beaker and water	g	
6. Volume of water added to the 50-mL beaker	mL	
7. Mass of water added to the 50-mL beaker	g	
8. Density of water	g/mL	

Experiment 1   Safety and Laboratory Techniques

Worksheet 1

Name _____

Date _____ Lab Instructor/Section _____

Safety in the laboratory Exercise

The following exercise should be attempted after you have read the introduction to this experiment and watched the safety video.

1. What non-clothing item must be worn in the laboratory at all times?

2. If the fire alarm goes off what do you do?

3. What is the proper procedure in the event of an acid spill?

4. How do you handle broken glass?

Experiment 1   Safety and Laboratory Techniques

In the space provided below draw a diagram of your laboratory noting all exits and safety equipment. Include the student benchtops (for perspective), fire extinguishers, eyewash stations, safety showers, exits, and broken glass containers.

SAFETY POLICIES

Safe practice in the chemical laboratory is a mutual responsibility and requires the full co-operation of everyone concerned at all times. This cooperation means that each student and instructor will observe safety precautions and procedures. The following general safety rules will be rigidly and impartially enforced throughout the semester. Noncompliance may result in dismissal from the lab and/or may result in a grading penalty.

1. Appropriate safety glasses must be worn at all times anywhere in the laboratory, even when not performing an experiment. Contact lenses should not be worn during the lab period.

2. Footwear should provide adequate protection against possible safety hazards (broken glass, reagent spills, etc.)

3. Food or drink will not be allowed in the laboratory.

4. Horseplay or other acts of carelessness are prohibited.

5. Unauthorized experiments are not permitted. Unapproved variations in experiments, including changes in the quantities of reagents, may be dangerous.

6. Every student is responsible for keeping his work area neat and orderly. After the experiment is over, clean the equipment and store it away correctly before you leave.

7. The instructor should be informed immediately of any safety hazards or accidents.

All accidents have causes and therefore can be prevented. Pay careful attention to what you are doing in each experiment, follow all instructions and use common sense. Be aware of what your neighbors do – you may be a victim of their accidents. Do not hesitate to comment tactfully to a neighbor whom you observe engaging in an unsafe practice. Thoroughly acquaint yourself with the location and use of emergency equipment (fire extinguishers, eye-wash stations, showers, etc.) around the lab. With the positive approach of good safety practice, all personal injuries can be avoided.

I have read and understood the safety rules outlined above. I agree to abide by them at all times while participation in Chemistry _____ laboratory, Section _____.

(Signed)_____ (Date)_____

Teaching Assistants Name_____

EMERGENCY & HEALTH INFORMATION

Experiment 1   Safety and Laboratory Techniques

Please print.

Your Name _____

Campus Phone_____

Campus or Home address_____

_____

_____

Person to be contacted in case of an accident:

Name:_____ ___

Phone:_____ ___

Do you have any health problems or disabilities that may cause difficulties in the Chemistry laboratory? If so, please describe them briefly below. This information will remain confidential.

_____
_____
_____
_____
_____

Contact lenses in Chemical Laboratories

The following is taken from "Handbook of Laboratory Safety" N.V. Stoek, Ed, 2nd ed., Chemical Rubber Co., Cleveland OH, 1971: "Contact lenses worn by persons working in laboratories can increase injury from chemical splashes because the wearer may not be able to remove the lenses to permit thorough irrigation, and a person giving first aid may not know that contact lenses are being worn or how to remove them. It is recommended that contact lenses not be worn in laboratories in which chemicals are handled or that wearers be sure to use full eye protection at all times."

In the pamphlet, "Use of Contact Lenses in Industry," published by the Council on Occupational Health of the American Medical Association, there are three paragraphs, which are particularly applicable to wearing contact lenses in laboratories.

"Many physicians believe that the substitution of contact lenses for spectacles in industrial workers is contraindicated in workers whose eyes may be exposed to dusts, molten metals, or irritant chemicals. Small foreign bodies, which normally are washed away by tears, sometimes become lodged beneath contact lenses where they may cause injury to the cornea. Similarly, chemicals splashed into the eye may be trapped under a contact lens and cause extensive corneal damage before the lens can be removed and the eye adequately irrigated."

"For effective protection for the eyes, the contact lens wearer should use in addition to his contact lenses the same approved face shields. Conventional safety spectacles, or goggles for protection against job hazards, as would any other worker on a similar job. Since removal of a contact lens for urgent irrigation after injury is made is so difficult by spasms of the eyelids, the contact lens wearer is in even greater need of these protections than his or her counterpart who does not wear contact lenses, if the job carries high potential risk of eye injury."

"Contact lenses are not in themselves protective devices in fact may increase the degree of injury to the eyes. Contact lens wearers in similar employment should wear the same eye-protective device used by other workers."

I have been informed of and understand the hazards associated with wearing contact lenses in the laboratory. I agree to wear safety goggles at all times while participating in Chemistry _____ laboratory.

Signed _____ Date _____

Experiment 2

Physical Properties of Substances

Objective

To measure the melting point, boiling point and density of several substances. To determine the identity of an unknown substance based on its measured density.

Equipment and Chemicals

100-mL graduated cylinder
150 and 250-mL beaker
Melting point tube – (side shelf)
Thermometer
One-hole rubber stopper with slit
Buret

Phenyl benzoate
Naphthalene
p-Dichlorobenzene
Unknown liquids
Unknown metals
Saltwater solution

Safety Precautions

Wear approved eye protection
Avoid inhaling chemicals. Be careful with open flames.

Experiment 2 — Physical Properties of Substances

Principles

Chemists measure the physical properties of substances, properties that can be measured without changing the sample, in order to distinguish one substance from another and to determine how the substance might be useful for some practical application. Some of the more important physical properties are briefly defined below.

Boiling point: The temperature at which the vapor pressure of a liquid is equal to the external or atmospheric pressure.

Melting point: The temperature at which solid and liquid are in equilibrium (one can see both solid and liquid in close contact). Melting points of pure substances do not change greatly with pressure and generally occur over a small temperature range. The melting point of an impure substance is generally lower than the pure substance and occurs over a large temperature range.

Density: The mass of a substance that occupies 1.0 mL (liquid) or 1.0 cm^3 (solid). The density of a solid can be determined by observing the volume of water that is displaced when a known mass of the solid is submerged in it.

Example: An unknown solid that weighs 127.4 grams is submerged into water. The initial volume of the water is 1,235.0 mL and a final volume is 1,268.5 mL. Calculate the density of the solid.

Density has units of g/cm^3 or g/mL. The mass of the solid is 127.4 grams and its volume is equal to the volume of displace water:

Displaced water =
1,268.5 − 1,235.0 mL = 33.5 mL

The density of the solid is
$$\frac{127.4 \text{ grams}}{33.5 \text{ mL}} = 3.80 \text{ g/mL or } 3.80 \text{ g/cm}^3$$

Procedure

Melting Point (A)

The melting point is usually determined on a very small sample of solid which has been placed in a thin tube closed at one end called a melting point tube. These tubes are available on the side shelf. Place less than 0.1 g of one of the solids on the side shelf on a piece of weighing paper. Fill a melting point tube by tapping the open end (the top) into the solid, inverting the tube and gently tapping the bottom of the tube on the bench top. Repeat if necessary until about ¼ of an inch of solid is at the bottom of the tube. You may select any of these solids for this experiment:

1. Phenyl Benzoate
2. Naphthalene
3. p-Dichlorobenzene

After filling the melting point tube, fasten it to a thermometer using a rubber circle or band, which you may obtain from the side shelf. The solid in the melting point tube must be adjacent to the thermometer bulb. Place the thermometer into a one-hole split stopper (caution: slide the thermometer gently into the stopper), fill the 250-mL beaker about half full with the salt water solution found on the side shelf and assemble the thermometer and beaker of saltwater as shown in **Figure 1**.

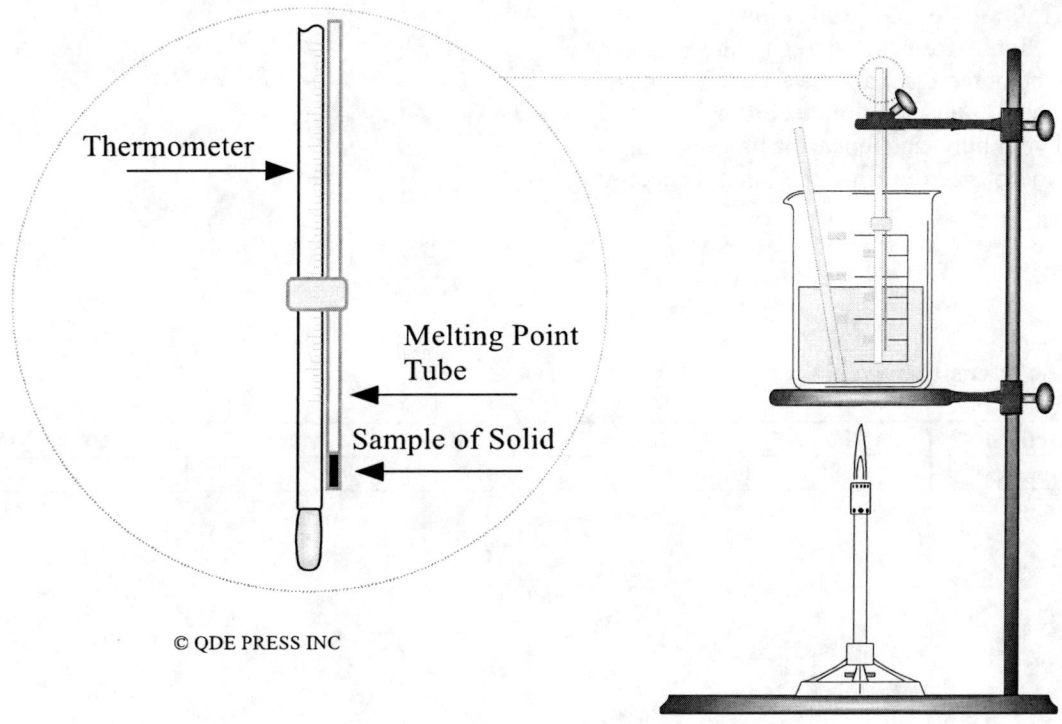

Figure 1

Heat the water with continuous stirring. Watch the solid carefully and note the temperature which it melts. The most accurate value for the melting point is obtained when the temperature is raised very slowly. Note: Do not throw the saltwater solution away, you will use it for the next experiment.

Boiling Point (B)
Continue heating the solution from Part A until it comes to a gentle boil. Read the temperature to the nearest 0.1 °C. Once the temperature remains constant for 15 – 20 seconds, record it in the Data Section.

Density of a Solid (C)
Use a beaker to collect an unknown metal from your instructor. Get enough metal to have about 100 g of the metal. The metal will be one of those listed in the density table below. Determine the mass of a 150 mL beaker to the nearest 0.01 gram. Add the metal and determine the mass of the beaker and metal. Record both masses in the Data Section. Fill a 100 mL graduated cylinder about half full with tap water and read the volume accurately to 0.1 mL (record in the Data Section). Add the metal and record the new volume level of the water. Pour out the water, dry the metal and return the metal to its beaker.

Density of a Liquid (D)

To calculate the density of an unknown solution, we need to measure the mass of a measured volume of solution. A 50 mL buret will be used to measure the volume. Clean and rinse your buret and allow it to drain. Obtain about 150 mL of the unknown solution. Rinse your buret twice with 5 to 10 mL of this solution and drain. Fill the buret to about the zero mark and drain out some liquid so that no air bubbles are in the tip and the solution is now slightly below the zero mark. Read the buret accurately to two decimal places and record the data. Now determine the mass of a dry 150-mL beaker. Add around 30-35 mL of the liquid from the buret to the beaker. Read the buret again and determine the mass of the beaker and solution, recording all data. Refill your buret and repeat the process again. You do not need to rinse the buret a second time.

Table of Density in g/cm^3

Element	Fe	Zn	Cd	Sn	Pb
Density	7.85	7.10	8.65	7.28	11.4

Experiment 2   Physical Properties of Substances

Worksheet 2

Name _____

Date _____   Lab Instructor/Section _____

Pre-lab	____/20
Data	____/20
Post-lab	____/20
Safety/Part.	____/40
Total	____/100

Data Sheet

A. Name of Compound tested   _____

   Melting point   _____ °C

B. Boiling point of salt-water solution   _____ °C

C. Density of a Solid
   Sample letter_____

   1. Mass of empty beaker   _____ g
   2. Mass of beaker + metal   _____ g
   3. Mass of metal   _____ g
   4. Volume of water   _____ mL
   5. Volume of water + metal   _____ mL
   6. Volume of solid   _____ mL
   7. Density of Solid [(3)/(6)]   _____ g/mL or g/cm^3
   8. Name of solid (see table)   _____

D. Density of a Liquid

		First Run	Second Run
1.	Initial buret reading	_____ mL	_____ mL
2.	Final buret reading	_____ mL	_____ mL
3.	Volume of liquid	_____ mL	_____ mL
4.	Mass of beaker	_____ g	_____ g
5.	Mass of beaker + liquid	_____ g	_____ g
6.	Mass of liquid	_____ g	_____ g
7.	Density of liquid [(6)/(3)]	_____ g/mL	_____ g/mL

- 15 -

Questions

1. What error, if any, would the calculated density of the solid have if the material had a hollow center?

2. Why is it important to constantly stir the salt-water solution while measuring the melting point of the solid?

Experiment 2 Physical Properties of Substances

Pre-Laboratory 2

Name _____

Date _____ Lab Instructor/Section _____

Pre-laboratory

1. Calculate the density of an object that has a dry mass of 12.7 g and displaces 1.57 mL of water when submerged. (Show all work)

2. From the table of densities found in this experiment, iron has a density of 7.85 g/cm^3 and water has a density of 1.0 g/mL. Iron is denser than water and will sink when placed in water. Explain how ships made from iron stay afloat using density principles.

3. What mass of solid lead would displace exactly 234.6 liters of water? (hint: use the table of densities)

Experiment 3

Percent of Copper in Copper(II) Sulfate Pentahydrate

Objective

To measure the amount of copper in a copper compound and calculate the experimental and theoretical percents of copper found in the compound.

Equipment and Chemicals

$CuSO_4 \cdot 5H_2O$	Acetone	1.0 M HCl
1.0 M H_2SO_4	250-mL beaker	
Zinc (granulated)	50-mL beaker	

Safety Precautions

Wear approved eye protection. Be careful to have no flames present in the laboratory after the zinc is added to the reaction mixture. The H_2 gas produced in this step can react violently with O_2 in the air.

Principles

During this lab, you will experimentally determine the mass percent of the element copper (Cu) in the compound copper(II) sulfate pentahydrate ($CuSO_4 \cdot 5H_2O$). Hydrated compounds have a specific number of water molecules per formula unit of ionic compound. The water molecules are held in place by forces presented at the end of CHEM 111. The naming involves a prefix (here penta- stands for 5) with the word hydrate for H_2O. The waters of hydration, as they are called, are included in the formula and molar mass. For example, the molar mass of $Ba(OH)_2 \cdot 8H_2O$ would be correctly calculated to be 1(Ba) + 10(O) + 18(H) = 1(137.33) + 10(16.00) + 18(1.008) = 315.47 u for formula mass or 315.47 g/mol for molar mass. The mass percent of an element is calculated by (mass element/mass compound) x 100%. For this experiment, you will need to experimentally determine the mass of copper in a sample of copper(II) sulfate pentahydrate. After you finish the experiment, you may compare your result to the theoretical mass percent calculated from the atomic mass of Cu and the formula mass of $CuSO_4 \cdot 5H_2O$.

The percent by mass of any part of a substance is the mass of that part divided by the mass of the whole substance, times 100%. In order to experimentally obtain the mass of any part of a substance it must be separated from the whole. In this experiment, the reaction shown below where zinc is exchanged for copper will be used to separate the copper from a measured mass of copper(II) sulfate pentahydrate. The abbreviations (aq) and (s) stand for aqueous solution (a solution where water is the solvent) and solid, respectively.

$Zn(s) + CuSO_4(aq) \rightarrow Cu(s) + ZnSO_4(aq)$

You will add zinc metal (black/gray solid) to a blue solution of Cu^{2+} causing an oxidation-reduction (redox) reaction that produces the "copper colored" Cu(s) and a colorless solution of Zn^{2+} (the SO_4^{2-} is a "spectator" ion that does not undergo change in the reaction). The procedure is successful because the $ZnSO_4$ is soluble and the copper solid can be separated from the other components by decantation, pouring off the liquid into another container. An excess of zinc metal is used in the experiment to completely convert all of the copper ions to the metal in a short time period. The excess zinc can be removed before weighing the copper because it reacts with dilute acid as shown below for hydrochloric acid.

$Zn(s) + 2HCl(aq) \rightarrow ZnCl_2(aq) + H_2(g)$

Copper does not react under similar conditions and can be collected after the zinc is removed.

The percent copper in the sample is determined by dividing the mass of copper collected in the experiment by the mass of the copper compound. For example, in an experiment similar to the one above starting with $CuCl_2$, if the mass of the copper compound was measured as 2.32 g and the mass of the copper collected measured as 1.09 g, the percent of copper would be:

Mass percent copper =

$$\frac{1.09 \text{ g}}{2.32 \text{ g}} \times 100\% = 47.0\%$$

In this case we know the formula of the starting copper compound and can calculate what the theoretical value should be. The molar mass of Cu is 63.55 g/mol and that of $CuCl_2$ is 134.45 g/mol.

Theoretical mass percent copper =

$$\frac{63.55 \text{ g/mol}}{134.45 \text{ g/mol}} \times 100\% = 47.27\%$$

This value is slightly higher than the calculated value, indicating that in the actual experiment, not all of the copper was collected

Procedure

Determine the mass of an empty 250-mL beaker. Transfer about 1.5 grams of the $CuSO_4 \bullet 5H_2O$ into the 250-mL beaker and then determine the mass of the beaker and its contents. Enter the masses in the data section. Subtraction of the second mass of the 250-mL beaker from the first yields the mass of $CuSO_4 \bullet 5H_2O$ transferred to the 250 mL beaker. This method is known as *weighing by difference* and is generally the preferred method.

Add 50 mL of 1 M H_2SO_4 to the 250-mL beaker, then warm gently and stir the mixture until the sample is dissolved. **Turn off the flame** (the gas evolved in the next step is hydrogen) - no open flames should be near your reaction at this point because the reaction of H_2 with O_2 to make H_2O is very violent and is set off by a spark or flame. Place about 1.2 grams (weighed to 0.1 g) of zinc metal in the solution of copper sulfate and cover the beaker with a watch glass. Allow the reaction to proceed, removing the cover every few minutes to stir the solution. The reaction is complete when the solution is colorless and no more gas is evolved. *If the solution is still blue add a tiny amount of zinc to react with the copper ions still in solution. If gas evolution still occurs when the solution is colorless, add 2 mL of dilute hydrochloric acid and stir and heat gently until gas evolution ceases.*

Allow the copper to settle to the bottom of the beaker and carefully decant off the liquid. Add 25 mL of deionized water to the beaker and stir vigorously for a few minutes. Again allow the solid to settle to the bottom of the beaker and carefully decant off the liquid. Wash the metal in this manner once more with 25 mL of water and finally two times with 5 mL of acetone. After carefully pouring off the last of the acetone, the residual acetone may be evaporated by stirring the copper solid for a few minutes. Determine the mass of the copper and beaker and record the value in the data section. Discard the copper into a beaker on the side shelf labeled "Copper Waste". Perform the calculations.

Note: As in any experiment of this type, some of the final product (copper in this case) will be lost before determining the mass. Your actual yield might be lower than the yield predicted (theoretical yield) from the masses of the starting materials (the limiting reagent). Thus, your percent copper as found in the experiment may be lower than the theoretical amount. It cannot be higher than the theoretical amount if you weighed pure, dry copper at the end and your initial mass of $CuSO_4 \cdot 5H_2O$ was correct

Worksheet 3

Name _____

Date _____ Lab Instructor/Section _____

Pre-lab	____/20
Data	____/20
Post-lab	____/20
Safety/Part.	____/40
Total	____/100

Data sheet

1. Mass of the empty 250-mL beaker _____ g

2. Mass of 250-mL beaker + sample _____ g

3. Mass of sample added to 250-mL beaker _____ g

4. Mass of 250-mL beaker + copper _____ g

5. Mass of copper _____ g

6. Experimental mass percent of copper in sample _____ %

7. Calculated theoretical mass percent of copper in $CuSO_4 \bullet 5H_2O$ _____ %

Calculations:

Don't forget your post-lab question on the back of this page!

Questions

1. Your actual experimental percent of copper is probably different from the theoretical value. Discuss possible reasons to account for the difference.

Pre-Laboratory 3

Name _____

Date _____ Lab Instructor/Section _____

Pre-laboratory

Barium metal was quantitatively precipitated from a 1.52 g sample of $BaCl_2 \cdot 2H_2O$. The mass of the barium that was collected was 0.844 g.

1. Calculate the experimental mass percent of barium in the sample.

2. Calculate the theoretical mass percent of barium in $BaCl_2 \cdot 2H_2O$.

Experiment 4

Limiting Reactant and Percent Yield

Objective

To synthesize iron(III) hydroxide from iron(III) nitrate and sodium hydroxide, calculate the theoretical yield from the limiting reactant and determine the percent yield.

Equipment and Chemicals

Iron(III) nitrate nonahydrate, $Fe(NO_3)_3 \cdot 9H_2O$
Sodium hydroxide, NaOH
400 ml beaker
Two 100 ml beakers
Deionized water

Buchner funnel assembly
5.5 cm filter paper
Two stirring rods
Spatula

Safety Precautions

Wear approved eye protection. Do not handle the reactants or product, the hydroxides are strong bases and can burn your skin. If you touch them wash your hands immediately with water. Wear gloves.

Principles

The major function of many chemists is the preparation of new chemical compounds or the development of new and improved methods for preparing known compounds. In order to develop the experimental techniques needed to carry out this type of chemistry, we will prepare a variety of known compounds in this introductory chemistry laboratory.

This experiment uses the synthesis of iron(III) hydroxide as an example of a typical chemical reaction. This preparation is successful because both of the reactants and one of the products, sodium nitrate, are soluble in water, whereas the desired product, iron(III) hydroxide, is not very soluble. Thus, the desired product can be isolated by a filtration of the reaction solution.

$$Fe(NO_3)_3(aq) + 3NaOH(aq) \rightarrow Fe(OH)_3(s) + 3NaNO_3(aq)$$

In this experiment, we will weigh out the reactants as solids. The solids will be dissolved in water and these solutions mixed to carry out the chemical reaction, From the mass of the reactants we can determine the maximum quantity of $Fe(OH)_3$ product that can be obtained in the reaction. This maximum amount is known as the *theoretical yield*. As both reactants are to be weighed out, to calculate the theoretical yield the *limiting reactant*, the reactant that is completely consumed when a chemical reaction occurs, must be determined. After the reaction is over and the solid product is isolated, it will be weighed to determine how close the *actual yield*, the amount of product isolated in the experiment, is to the theoretical yield (it cannot be greater). Chemists express the success in this area as a *percent yield* where

Percent yield =

$$\frac{\text{actual yield}}{\text{theoretical yield}} \times 100\%$$

For example, in a related experiment, the synthesis of magnesium hydroxide from magnesium chloride hexahydrate and sodium hydroxide, 2.01 g of magnesium chloride hexahydrate and 1.32 g of sodium hydroxide are weighed out, dissolved in water, these solutions mixed and 0.44 g of solid magnesium hydroxide isolated. In order to determine the limiting reactant, the theoretical yield and the percent yield, the chemical equation is needed.

$MgCl_2(aq) + 2NaOH(aq) \rightarrow$

$\qquad Mg(OH)_2(s) + 2NaCl(aq)$

To determine the limiting reactant, the grams of reactants need to be converted to moles of reactants using the correct moles-mass conversion factor. For magnesium chloride, in the molar mass of magnesium chloride hexahydrate we need to include the mass of the six water molecules because they are present in the solid that was weighed out – the molar mass of the hexahydrate is 203.3 g/mol. The 6 water molecules were not written in the equation above because the reaction is carried out in water, and they just become part of the solvent. The moles of $MgCl_2$ are the same as moles of $MgCl_2 \cdot 6H_2O$

Amount $MgCl_2$ =

$2.01 \text{g } MgCl_2 \cdot 6H_2O \left(\dfrac{1 \text{ mol } MgCl_2 \cdot 6H_2O}{203.3 \text{ g } MgCl_2 \cdot 6H_2O} \right) =$

$0.00989 \text{ mol } MgCl_2 \cdot 6H_2O =$

$0.00989 \; MgCl_2$

Amount NaOH = 1.32 g NaOH

$\left(\dfrac{1 \text{ mol NaOH}}{40.00 \text{ g NaOH}} \right) = 0.0330 \text{ mol NaOH}$

The molar amounts of each reactant need to be converted to moles of magnesium hydroxide using the coefficients of the equation to determine the limiting reactant.

Amount $Mg(OH)_2$ based on $MgCl_2$ =

$0.00989 \text{ mol } MgCl_2 \left(\dfrac{1 \text{ mol } Mg(OH)_2}{1 \text{ mol } MgCl_2} \right) =$

$0.00989 \text{ mol } Mg(OH)_2$

Amount $Mg(OH)_2$ based on NaOH =

0.0330 mol NaOH $\left(\dfrac{1 \text{ mol Mg(OH)}_2}{2 \text{ mol NaOH}}\right) =$

0.0165 mol Mg(OH)$_2$

The magnesium chloride is the limiting reactant (less product is produced from it) and along with the molar mass of Mg(OH)$_2$ is used to calculate the theoretical yield.

Mass Mg(OH)$_2$ =

0.00989 mol Mg(OH)$_2$ $\left(\dfrac{58.33 \text{ g Mg(OH)}_2}{1 \text{ mol Mg(OH)}_2}\right) =$

0.577 g Mg(OH)$_2$

Use this theoretical yield and the actual yield obtained in the experiment to calculate the percent yield.

Percent yield = $\dfrac{\text{actual yield}}{\text{theoretical yield}}$ x 100%

= $\dfrac{0.44 \text{ g}}{0.577 \text{ g}}$ x 100% = 76.%

Procedure

Weigh out accurately about 1.9 g Fe(NO$_3$)$_3$·9H$_2$O and about 0.75 g of NaOH. Record the amounts in the Data Section. Remember as in the above example in the calculation that converts the number of grams to moles for Fe(NO$_3$)$_3$·9H$_2$O that the nine water molecules must be included in the molar mass needed to convert grams to moles (molar mass of Fe(NO$_3$)$_3$·9H$_2$O = 404.1 g/mol). Place each solid in a separate 100 mL beaker, add 20 mL of deionized water to each and stir each with a glass rod to dissolve the solids. Pour the contents of one beaker into the other, a dark solid should form immediately. Stir this now heterogeneous mixture, then allow it to settle for at least 5 minutes.

Collect your product using a Buchner funnel placed in a suction flask (see drawing) attached to an aspirator or vacuum pump. *Note that it is imperative to clamp the suction flask so that it will not tip over.*

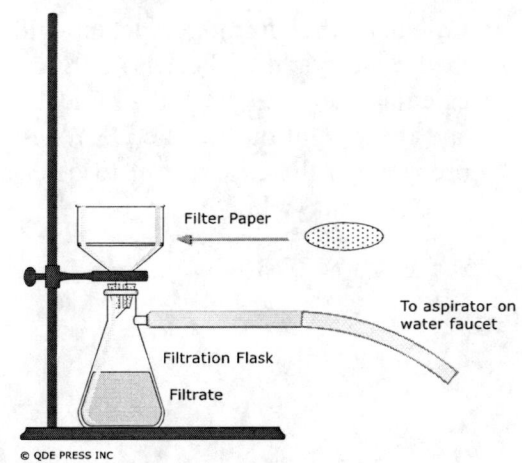

Separation Technique: Suction Filtration

Wash the solid twice with cold water to remove any of the liquid filtrate from which the solid precipitated; this filtrate probably contains unwanted impurities. To wash the solid, disconnect the aspirator/pump hose, add enough cold water to completely cover the solid, gently (do not disturb the filter paper) stir the solid with the glass rod so as to ensure good contact between the water and the solid, and finally reconnect the aspirator/pump and suck to dryness. Do this procedure twice, leaving the aspirator/pump on longer the second time to help dry the solid. Let the solid sit for a few minutes with the aspirator/pump on. Weigh a 400 mL beaker, record the weight in the Data Section and transfer the solid and filter paper to it. This transfer is accomplished by first loosening the circumference of the cake with a spatula and inverting the funnel over the beaker. To complete the drying, turn on a hot plate and set the heat dial half-way between off and maximum. Place the 400 mL beaker on it for about 7 minutes. Remove the filter paper from the beaker and scrape any product remaining on it back into the beaker. After 3 more minutes heating, weigh the beaker and product, record this weight.

Calculate the limiting reactant and the theoretical yield of $Fe(OH)_3$. Show the calculations. Use the theoretical yield and the weight of your $Fe(OH)_3$ actually prepared in the experiment to determine your percent yield.

Worksheet 4

Name _____

Date _____ Lab Instructor/Section _____

Pre-lab	____/20
Data	____/20
Post-lab	____/20
Safety/Part.	____/40
Total	____/100

Data sheet

1. Mass of $Fe(NO_3)_3 \cdot 9H_2O$ _____ g

2. Mass of NaOH _____ g

3. Amount $Fe(OH)_3$ based on $Fe(NO_3)_3 \cdot 9H_2O$ _____ mol

4. Amount $Fe(OH)_3$ based on NaOH _____ mol

5. Theoretical yield of $Fe(OH)_3$ in grams _____ g

6. Mass of 400 mL beaker _____ g

7. Mass of beaker and $Fe(OH)_3$ _____ g

8. Mass of $Fe(OH)_3$ (actual yield of reaction) _____ g

9. Percent yield _____ %

Explain why your yield cannot exceed 100%. If your yield in the actual experiment did exceeded 100%, explain how this happen, given the previous sentence.

Pre-Laboratory 4

Name _____

Date _____ Lab Instructor/Section _____

1. What is the theoretical yield in grams of Ca(OH)₂ if 2.04 g of KOH reacts with 3.02 g of CaCl₂·2H₂O as shown by the below equation?

$$CaCl_2(aq) + 2KOH(aq) \rightarrow Ca(OH)_2(s) + 2KCl(aq)$$

In the part of the calculation that converts the number of grams to moles for CaCl₂·2H₂O, the two water molecules must be included in the molar mass needed to convert grams to moles (molar mass of CaCl₂·2H₂O is 147.0 g/mol). What is the limiting reactant?

Limiting reactant _____   Theoretical yield _____ g

2. You carry out the synthesis of Ca(OH)₂ using the 2.04 g of KOH and 3.02 g of CaCl₂·2H₂O as in #1, and your yield of dry Ca(OH)₂ is 1.22 g. What is the percent yield for the reaction?

Percent yield _____ %

Experiment 5

Acid-Base Titration: Determination of the Concentration of a NaOH Solution

Objective

To determine the concentration of an unknown solution of sodium hydroxide by titration, a volumetric method of analysis, using the known concentration of an oxalic acid solution, which is prepared by measuring the mass of oxalic acid solid and total volume of an aqueous solution of that solid.

Equipment

Solid oxalic acid, $H_2C_2O_4$
100-mL volumetric flask with stopper
NaOH solution, unknown concentration
Phenolphthalein
Stirring bar

Two 50-mL burets
Weighing paper
Deionized water
Stirring plate
250-mL Erlenmeyer flask

Safety Precautions

Wear safety glasses. Handle the sodium hydroxide solution with care, it is caustic. Avoid spilling it on your skin or clothing. If your hands feel "soapy" that is from the sodium hydroxide solution: rinse thoroughly with water in the sink.

Principles

A titration is a process used to determine the quantity of one substance by adding a measured amount of a second substance. In this experiment, you will titrate a sodium hydroxide (NaOH) solution of unknown concentration with an oxalic acid ($H_2C_2O_4$ Figure 1) solution for which the molarity will be known from the mass of oxalic acid added to a solution of known volume. The equation for the reaction in this experiment is:

$2NaOH(aq) + H_2C_2O_4(aq) \rightarrow$

$Na_2C_2O_4(aq) + 2 H_2O(\ell)$ (1)

Note from equation 1 that it takes two moles of the NaOH to react with one mole of $H_2C_2O_4$.

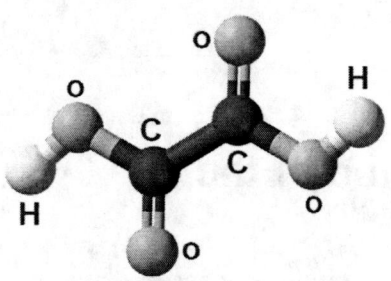

Figure 1. Oxalic Acid

The most common unit of concentration is molarity, M, which is defined as the number of moles of solute present in one liter of solution:

$$\text{molarity} = M = \frac{\text{moles of solute}}{\text{liters of solution}} \quad (2)$$

Molarity is a convenient concentration unit because the number of moles of solute in a solution is needed in stoichiometric calculations. Molar concentrations relate the volume of solution (expressed in liters) to the number of moles of solute present. For example, each liter of a 1.20 M NaOH solution will contain 1.20 moles of NaOH.

1 L NaOH soln contains 1.20 mol NaOH

Equivalencies of this type can be used to convert volume of solution (in L) to number of moles or *vice versa*.

For example, a 750-mL solution that is 1.20 M in NaOH contains:

$$0.750 \text{ L NaOH soln} \left(\frac{1.20 \text{ mol NaOH}}{1 \text{ L NaOH soln}}\right)$$
$$= 0.900 \text{ moles of NaOH} \quad (3)$$

In this experiment you will determine the number of moles of NaOH contained in a known volume of NaOH solution. The concentration of the solution can then be calculated by dividing the number of moles of NaOH by the volume of solution. For example if it were determined that a 100-mL solution of NaOH contained 0.35 moles of NaOH the concentration would be:

$$\text{molarity of NaOH} = \left(\frac{\text{moles of NaOH}}{\text{volume of NaOH soln.}}\right) =$$

$$\left(\frac{0.35 \text{ mol of NaOH}}{0.100 \text{ L of NaOH soln.}}\right) = 3.5 \, M \text{ NaOH} \quad (4)$$

Therefore, the key is determining the number of moles of NaOH contained in a known volume of NaOH solution. From Equation 1 we observe that one mole of $H_2C_2O_4$ reacts with exactly *two* moles NaOH producing one mole of $Na_2C_2O_4$ and two moles of water. The reaction between an acid and a base producing a salt and water is called **neutralization.** When enough NaOH (which supplies one mole of OH^- per mole of NaOH) is added to the solution of $H_2C_2O_4$ (which supplies *two* moles of H^+ per mole of $H_2C_2O_4$), the solution is neutralized. The point at which the stoichiometrically equivalent amounts of the two reactants are present is called the **equivalence point**. At the equivalence point the reaction stoichiometry can be used to calculate the amount of NaOH in the sample from the measured volume and known concentration of $H_2C_2O_4$ solution. For example, what is the concentration of a NaOH solution if a 75-mL sample required 65.0 mL of a 2.50 M $H_2C_2O_4$ solution in order to reach the equivalence point?

As in Equation 3:

moles of $H_2C_2O_4$ =

$$0.0650 \text{ L } H_2C_2O_4 \text{ soln} \left(\frac{2.50 \text{ mol } H_2C_2O_4}{1 \text{ L of } H_2C_2O_4 \text{ soln}}\right)$$
$$= 0.162 \text{ mol of } H_2C_2O_4$$

From the stoichiometric relationship shown in Equation 1:

moles of NaOH =

$$0.162 \text{ mol } H_2C_2O_4 \left(\frac{2 \text{ mole of NaOH}_2}{1 \text{ mole of } H_2C_2O_4} \right)$$

$$= 0.324 \text{ mol of NaOH}$$

NOTE: the conversion factor in this step is derived from the coefficients of Equation 1.

Using the procedure shown in Equation 4 we can conclude:

molarity of NaOH soln =

$$\left(\frac{0.324 \text{ mol of NaOH}}{0.075 \text{ L of NaOH soln}} \right) = 4.4 \, M \text{ NaOH}$$

In this experiment, you will prepare a standard solution of oxalic acid of known molar concentration by weighing the solid oxalic acid and dissolving this solid into a aqueous solution of known volume. A measured volume of this oxalic acid solution will be titrated with the NaOH solution of unknown concentration. The equivalence point will be determined by a change in color of the phenolphthalein indicator, a compound that shows no color in acid solution but turns pink in basic solution. The color change is known as the **end point** in the titration, and the indicator must be selected properly so the end point is close to the **equivalence point**.

Procedure

A standard oxalic solution is first prepared which is later used to standardize a sodium hydroxide solution of unknown concentration. To prepare the standard oxalic acid, on the balance tare a piece of weighing paper and weigh out accurately (to 0.01 g) between 2.2 and 2.5 g of oxalic acid. Enter this mass in the data section.

Transfer the solid acid to a 100-mL volumetric flask. Extreme care must be used to insure all of the solid is transferred to the flask. Fill the flask a little over half full and rotate the flask until the acid has all dissolved – then add just enough water to bring the solution level even with the etched line on the neck of the flask (the bottom of the meniscus should be exactly even with the line). Stopper the flask, shake it well and calculate the molarity of the oxalic acid solution.

To standardize your unknown sodium hydroxide, the first important step is to clean and rinse your burets. Alconox detergent powder makes a good cleaning solution. When your burets are clean, they should be rinsed several times with water and allowed to drain.

Rinse the burets twice with 5 to 10 mL of the solution to be used in each one, i.e., unknown sodium hydroxide solution in one and oxalic acid solution in the other. Allow the burets to drain well between the rinsings. Be sure your burets are labeled or marked to indicate the appropriate solution in each.

Fill the burets with the solution to a point above the zero mark and allow sufficient solution to run out so that the tips of the burets contain no air bubbles and the level of the solution is at or slightly below zero mark. Read your burets to an accuracy of two decimal places and enter the values in the data section. Remember to always read at the bottom of the meniscus. You are now ready to begin the titration to determine the molarity of your unknown sodium hydroxide solution.

Clean a 250-mL Erlenmeyer flask thoroughly rinsing it several times with deionized water. Make an initial reading of the volume of the oxalic acid solution buret, record the number, and add about 20 mL to the Erlenmeyer flask from the buret. Again read and record the volume in the buret. Add two drops of phenolphthalein indicator to the solution in the flask.

Add a stirring bar to the flask, place it on a stirring plate (cover the plate with a piece of white paper for better color contrast) and very gently stir. Read the initial volume of the sodium hydroxide buret. Place the flask under the tip of this buret and start titrating by adding the sodium hydroxide solution. At first, you may add the solution rapidly, but as the pink color tends to persist longer, decrease the rate of addition until finally you are adding individual drops. When one drop changes the color of the solution to pink which persists for *twenty seconds* or more, you have reached the "end point". If you should pass the end point by mistake, (indicated by a red color) you may add a little bit more oxalic acid from the buret until the solution with the indicator is colorless again. Now again titrate with the sodium hydroxide from your buret until the end point is reached. Obtain the final readings from both of your burets and record them into the data section.

Repeat this titration and calculate the molarity of your unknown sodium hydroxide solution using the data from each of the first two trials. If your results differ by more than 1%, a third trial should be made.

Experiment 5   Acid-Base Titration NaOH Solution

Worksheet 5

Name _____

Date _____ Lab Instructor/Section _____

Pre-lab	____/20
Data	____/20
Post-lab	____/20
Safety/Part.	____/40
Total	____/100

Part I. Oxalic Acid Solution Preparation

1. Mass of oxalic acid   _____ grams

2. Moles of oxalic acid   _____ moles

3. Molarity of oxalic acid solution   _____ moles/liter

Part II. Titration

1. Trial I   Oxalic Acid   Sodium Hydroxide

    Initial burette reading   _____   _____

    Final burette reading   _____   _____

    mL of solution   _____   _____

2. Trial II   Oxalic Acid   Sodium Hydroxide

    Initial burette reading   _____   _____

    Final burette reading   _____   _____

    mL of solution   _____   _____

3. Moles of oxalic acid transferred to Erlenmeyer flask

        Trial I _____   Trial II _____

4. Moles of sodium hydroxide calculated from moles of oxalic acid in Part II Step 3.

        Trial I _____   Trial II _____

5. Molarity of sodium hydroxide (NaOH) solution   I _____

                                                                                    II _____

                                                                                  Average _____

Questions

1. Is the calculated sodium hydroxide concentration higher, lower or the same as the actual sodium hydroxide concentration in the flask, if the oxalic acid that was weighed was not completely transferred to the volumetric flask? Explain your answer.

2. A 0.261 g sample of $NaHC_2O_4$ (one acidic proton) required 17.5 mL of sodium hydroxide solution for complete reaction. Write the equation for this reaction and determine the molar concentration of the sodium hydroxide solution.

Experiment 5   Acid-Base Titration NaOH Solution

Pre-Laboratory 5

Name _____

Date _____ Lab Instructor/Section _____

Pre-laboratory

1. Calculate the mass of oxalic acid ($H_2C_2O_4$) needed to make 100 mL of a 0.250 M solution.

2. How many moles of solute are present in 300. mL of a 0.60 M solution of NaOH?

3. What volume of a 0.50 M solution of H_2SO_4 is required to completely neutralize 3.0 grams of NaOH? Write the equation.

Experiment 6

Heat of Formation: Hess's Law

Objective

To determine the change in enthalpy of two reactions and use Hess's Law to calculate the standard heat of formation of magnesium oxide, MgO.

Equipment

CBL system
TI graphing calculator
AC adapter
Vernier temperature probe
Ring stand
Utility clamp
1.0 M HCl
Slit stopper

100-mL graduated cylinder
Magnesium ribbon, Mg
Styrofoam cup
Magnesium oxide, MgO
Stirring rod
Balance
Weighing paper
250-mL beaker

Safety Precautions

Wear approved eye protection

HCl is corrosive and can damage skin eyes and clothing. Handle it with care. If skin contact occurs, wash immediately with copious amounts of water.

Avoid inhaling magnesium oxide dust.

Principles

All chemical reactions are accompanied by a change in energy. For example, natural gas is burned to heat homes and cook our food. The area of chemistry that deals with the heat evolved or gained during a chemical reaction is known as **thermochemistry**. This experiment shows how to measure heat changes in a chemical reaction and how to use this information to determine the heat changes associated with other, related reactions.

One unit of heat and other forms of energy is the calorie. A **calorie (cal)** is defined as the amount of heat required to raise the temperature of one gram of water by 1 degree Celsius. Another unit of heat and the one we will be using in this experiment is the **joule (J)**. There are 4.184 joules in one calorie. A **kilojoule (kJ)** is 1000 joules.

The process of measuring the amount of heat transferred in a chemical reaction is called **calorimetry**. One form of calorimetry is based on measuring the change in temperature of a known quantity of water. We must also know the **specific heat**, the quantity of heat required to raise the temperature of 1 gram of a substance (like water) 1 degree Celsius, having the units (J/g °C). The specific heat of water is 4.184 (J/g °C). If we know the mass, m, of a substance and its specific heat, C_s, we can determine the amount of heat, q, entering or leaving the substance by measuring the temperature change ($\Delta T = T_{final} - T_{initial}$) before and after the heat is gained or lost.

$$q_{soln} = mC_s\Delta T \qquad (1)$$

For example, if during an experiment the temperature of 500 grams of water increases from 20.0 °C to 85.0 °C how much heat (in joules) did the water absorb?

q = mass x specific heat x
            change in temperature

q = 500 g x 4.184 J/g°C x (85.0 °C – 20.0 °C) = 1.36 x 10⁵ J or 136 kJ

When an energy change occurs under constant pressure the heat absorbed by the system during the change is called the change in enthalpy of the reaction and given the symbol **ΔH**.

* If the chemical reaction releases heat (is exothermic) then ΔH is negative (ΔH <0)

* If the chemical reaction absorbs heat (is endothermic) then ΔH is positive (ΔH >0).

The techniques and equipment used in calorimetry depend on the nature of the process being studied. Many reactions occurring in solution are carried out at constant pressure, so that ΔH is measured directly. A very simple "coffee cup" calorimeter, as shown in **Figure 1**, is often used in general chemistry laboratories to illustrate the principles of calorimetry. Because the calorimeter is not sealed, the reaction occurs under the essentially constant pressure of the atmosphere. The heat gained or lost by the solution, q_{soln},

is readily calculated from the mass of the solution, its specific heat, and the temperature change, as seen in Equation 1. For dilute aqueous solutions, the specific heat of the solution will be approximately the same as that of water, 4.184 J/g °C. The heat gained or lost by the solution must be produced from the chemical reaction under study. Therefore, the enthalpy change of the reaction, ΔH_{rxn}, is equal in magnitude and opposite in sign from q_{soln}.

$$\Delta H_{rxn} = -q_{soln} \qquad (2)$$

A temperature increase (positive ΔT) of the water means that q_{soln} is positive and ΔH_{rxn} for the reaction is negative; the reaction is exothermic.

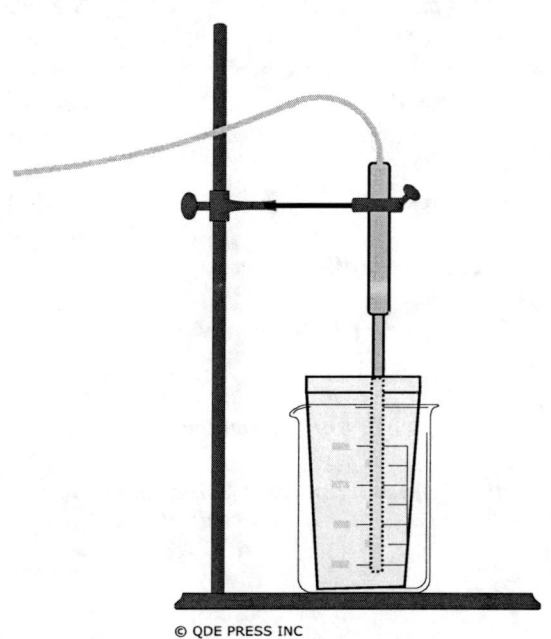

Figure 1

In addition to knowing that $\Delta H_{rxn} = -q_{soln}$, to convert a measured q value into ΔH_{rxn} we must remember that in thermochemical equations the ΔH_{rxn} value assumes that the reactants are present in mole amounts as shown by the coefficients. If we wanted to determine ΔH for reaction 3, we would need to study the reaction of one mole of $CH_4(g)$ with two moles of $O_2(g)$.

$$CH_4(g) + 2O_2(g) \rightarrow CO_2(g) + 2H_2O(\ell) \quad (3)$$

In the calorimetry experiments, much less than the mole amounts shown in the equations are actually used in the experiment. For example, in an experiment to determine ΔH for reaction 3 one might use only 1.00 g of CH_4 and determine q from the temperature rise and the amount of water present to be 55.4 kJ.

Remembering to reverse the sign, we have experimentally determined that:
1.00 g CH_4 = -55.4 kJ

The 1.00 g of CH_4 is equal to 0.0623 mol of CH_4, thus

0.0623 mol CH_4 = -55.4 kJ

In order to determine the ΔH for reaction 3, we must calculate ΔH for one mole of CH_4. **Hess's law** states that *the change in enthalpy*

$$\Delta H = 1 \, mol \, CH_4 \left(\frac{-55.4 \, kJ}{0.0623 \, mol \, CH_4} \right) = -889 \, kJ$$

for an equation obtained by adding two or more other equations is equal to the sum of the enthalpy changes for the added equations. We can calculate ΔH for any process if we can find equations, for which ΔH is known, which sum to the desired equation. For example, suppose you wanted to determine the ΔH for the following reaction.

$$C(s) + 1/2 \, O_2(g) \rightarrow CO(g) \quad (4)$$

It can be calculated if we know the enthalpy change for the reactions

$$C(s) + O_2(g) \rightarrow CO_2(g) \quad \Delta H = -393.5 \, kJ \quad (5)$$

$$CO(g) + \tfrac{1}{2} O_2(g) \rightarrow CO_2(g) \quad \Delta H = -283.0 \, kJ \quad (6)$$

In order to use equations 5 and 6, we arrange them so that $C(s)$ is on the reactant side and $CO(g)$ is on the product side of the arrow, as in the target reaction 4. Because equation 5 has $C(s)$ as a reactant, we can use that equation just as it is. Notice that the target reaction has $CO(g)$ as a product. Thus, we need to reverse equation 6 so that $CO(g)$ is a product. Remember that when reactions are reversed, the sign of ΔH is reversed. We arrange the two equations so that they can be added to give the target equation:

$$C(s) + O_2(g) \rightarrow CO_2(g) \quad \Delta H = -393.5 \, kJ$$
$$CO_2(g) \rightarrow CO(g) + \tfrac{1}{2} O_2(g) \quad \Delta H = 283.0 \, kJ$$
$$\overline{C(s) + \tfrac{1}{2} O_2(g) \rightarrow CO_{(g)} \quad \Delta H = -110.5 \, kJ}$$

An important method used for tabulating thermochemical data and using Hess's law is **enthalpy of formation**, the enthalpy change when one mole of a compound is formed from its constituent elements. This process is also called the *heat of formation* and is labeled ΔH_f, where the subscript *f* indicates that one mole of a substance is formed as shown below. Note that fractions are frequently used with reactants because the definition is to make *one mole of product*. Also, ΔH_f values can be either positive or negative.

$$C(s) + O_2(g) \rightarrow CO_2(g) \quad \Delta H_f = -393 \, kJ/mol$$

$$K(s) + 1/2 \, Cl_2(g) + 3/2 \, O_2(g) \rightarrow KClO_3(s)$$
$$\Delta H_f = -398 \, kJ/mol$$

$$2C(s) + H_2(g) \rightarrow C_2H_2(g)$$
$$\Delta H_f = 227 \, kJ/mol$$

In this experiment you will determine the heat of formation of magnesium oxide (MgO):

$$Mg(s) + \tfrac{1}{2} O_2(g) \rightarrow MgO(s) \quad \Delta H = ? \quad (7)$$

Experiment 6   **Heat of formation: Hess's Law**

This equation can be obtained by combining 8, 9 and 10:

$MgO(s) + 2HCl(aq) \rightarrow MgCl_2(aq) + H_2O(\ell)$ (8)

$Mg(s) + 2HCl(aq) \rightarrow MgCl_2(aq) + H_2(g)$ (9)

$H_2(g) + \frac{1}{2} O_2(g) \rightarrow H_2O(\ell)$ (10)

The pre-lab portion of this experiment requires you to combine equations 8, 9 and 10 to obtain equation 7 before you begin the lab. Using calorimetry, the heats of reaction for equations 8 and 9 will be determined. The ΔH for reaction 10 is -286 kJ.

Procedure

1. Wear safety glasses.
   - Work in pairs
   - Obtain a CBL unit from your TA.

2. Plug the temperature probe into Channel 1 of the CBL System. Use the link cable to connect the CBL System to the TI Graphing Calculator. Firmly press in the cable ends.

3. Turn on the CBL unit and the calculator. Press PRGM and select the CHEMBIO program. Press ENTER twice to proceed to the MAIN MENU.

4. Set up the calculator and CBL for one temperature probe and a temperature calibration.

   - Select SET UP PROBES from the MAIN MENU.
   - Enter "1" as the number of probes.
   - Select TEMPERATURE from the SELECT PROBE menu.
   - Enter "1" as the channel number.
   - Select USE STORED from the CALIBRATION menu.

5. Set up the calculator and CBL for data collection.

   - Select COLLECT DATA from the MAIN MENU.
   - Select TIME GRAPH from the DATA COLLECTION menu.
   - Enter "5" as the time between samples, in seconds.
   - Enter "96" as the number of samples (the CBL will collect data for a total of 8 minutes).
   - Press [ENTER]. Select USE TIME SETUP to continue. If you want to change the sample time or sample number, select MODIFY SETUP.
   - Enter "0" as the minimum temperature (Ymin).
   - Enter "50" as the maximum temperature (Ymax).
   - Enter "10" as the temperature increment (Yscl).
   - Do not start collecting data until instructed to do so in Step 9.

Reaction 1

6. Use a utility clamp and a slit stopper to suspend a temperature probe from a ring stand as shown in Figure 1.

7. Place a Styrofoam cup into a 250-mL beaker as shown in Figure 1. Measure out about 50.0 mL (record the mass to accuracy of your balance) of 1.00 M HCl into the Styrofoam cup. Lower the temperature probe into the solution. CAUTION: Handle the HCl solution with care. It can cause painful burns if it comes in contact with the skin.

8. Weigh out about 0.50g of magnesium oxide (record the mass to accuracy of your balance), MgO, on a piece of weighing paper. Record the exact mass used in your data table. CAUTION: Avoid inhaling magnesium oxide dust.

9. Press [ENTER] to begin data collection and obtain the initial temperature, T1. It may take several seconds for the temperature

probe to equilibrate at the temperature of the solution. After three or four readings at the same temperature (T1) have been obtained, add the white magnesium oxide powder to the solution. Use a stirring rod to stir the cup contents until a maximum temperature has been reached and the temperature starts to drop. Record the maximum temperature, T2. After 8.0 minutes, data collection is completed ("DONE" appears on the CBL screen).

10. Press ENTER to display a graph of temperature vs. time. To confirm the initial (T_1) and maximum (T_2) temperature values you recorded earlier, examine the data points along the curve. As you move the cursor right or left, the time (X) and temperature (Y) values of each data point are displayed below the graph.

11. Discard the solution as directed by your teacher.

Reaction 2

12. Press ENTER, then choose to repeat the data collection by selecting YES. Use the same Y-axis settings as in Part I. Repeat Steps 7-11 using about 0.25 g of magnesium ribbon (record the mass to accuracy of your balance) rather than magnesium oxide powder. The magnesium ribbon may have been pre-cut to the proper length by your teacher. **CAUTION:** *Do not breathe the vapors produced in the reaction!*

Data Analysis

1. In the spaces provided in the data sheet, calculate the change in temperature, ΔT, for Reactions 1 and 2 on the data table.

2. Calculate the heat released by each reaction, q, using the formula

   $q_{soln} = mC_s\Delta T$

   $C_S = 4.18$ J/g°C, and $m \approx 50.0$ g (use your data) of HCl solution. Convert joules to kJ in your final answer.

3. Determine ΔH. ($\Delta H = -q$)

4. Determine the moles of MgO and Mg used.

5. Use your Step 3 and Step 4 results (above) to calculate, as shown in the Introduction and in question 1 of the prelaboratory exercise, ΔH for Reactions 1 and 2 in the data table.

6. Determine ΔH_f for MgO(s). (Use your Step 5 results, your pre-lab work, and $\Delta H = -285.8$ kJ for Reaction 10).

7. Determine the percent error for the answer you obtained in Step 6. The accepted value for this reaction can be found in a table of standard heats of formation.

Experiment 6   Heat of formation: Hess's Law

Worksheet 6

Name _____

Date _____   Lab Instructor/Section _____

Pre-lab	____/20
Data	____/20
Post-lab	____/20
Safety/Part.	____/40
Total	____/100

Data sheet

	Reaction 1 (MgO)	Reaction 2 (Mg)
1. Mass of 1.00 M HCl (measure by difference)	_____ g	_____ g
2. Final temperature, T_2	_____ °C	_____ °C
3. Initial temperature, T_1	_____ °C	_____ °C
4. Change in temperature, ΔT	_____ °C	_____ °C
5. Mass of solid	_____ g	_____ g
6. Heat, q_{soln}	_____ kJ	_____ kJ
7. ΔH	_____ kJ	_____ kJ
8. Moles	_____ mol MgO	_____ mol Mg
9. ΔH_{rxn}	_____ kJ/mol	_____ kJ/mol
10. Determine ΔH_f for MgO, reaction 7. (8) _____ _____ (9) _____ _____ (10) _____ _____ (7) _____ _____	
		11. Percent error _____ %

Questions

1. Explain why a Styrofoam cup is used in this experiment.

2. Using your ΔH values for equations (7), (8) and (9) show that the ΔH for equation (10) is indeed −285.8 kJ. *Hint*: use Hess's law.

Experiment 6 Heat of formation: Hess's Law

Pre-Laboratory 6

Name _____

Date _____ Lab Instructor/Section _____

Pre-laboratory

1. A calorimetry experiment determined that the combustion of 1.01 g of $H_2(g)$ reacting with excess $O_2(g)$ gave q_{soln} = 143 kJ. Calculate ΔH for the reaction:

    $H_2(g) + \frac{1}{2} O_2(g) \rightarrow H_2O(\ell)$

2. In the space provided below, combine equations (8), (9), and (10) to obtain equation (7). It is not necessary to calculate ΔH.

    (8) _____

    (9) _____

    (10) _____

    (7) _____

Experiment 7

The Ideal Gas Law: Pressure-Temperature Relationship

Objective

Determine the affect a change in temperature has on the pressure of a gas held at a constant volume. This data will be used to express the relationship between the pressure and temperature of a confined gas in mathematical terms and confirm the value for absolute zero on the Celsius temperature scale.

Equipment

CBL system	125-mL Erlenmeyer flask
TI graphing calculator	Hot plate
AC adapter	Heavy wall plastic tubing
Vernier temperature probe	Slit stopper
Vernier pressure sensor	Four 1-liter beakers
1 Vernier adapter cable	Ice
Ring stand	1-hole stopper fitted with glass tube
Utility clamp	

Safety Precautions

Wear approved eye protection.
Boiling water can cause severe burns. Handle with care.

Principles

Experiments with a large number of gases reveal that four variables are needed to define the physical condition, or *state*, of a gas: temperature, T, pressure, P, volume, V, and the amount of gas, which is expressed as the number of moles, *n*. The equations that express the relationships among T, P, V, and *n* are known as the *gas laws*.

The Pressure-Volume Relationship

Boyle's law, which summarizes the relationship between pressure and volume sates that *the volume of a fixed quantity of gas maintained at constant temperature, is inversely proportional to the pressure.*

Inversely proportional means that when one measurement grows smaller the other becomes larger. Expressed in mathematical terms Boyle's law can be written as:

$$V = \text{constant} \times \frac{1}{P} \quad \text{or} \quad PV = \text{constant} \tag{1}$$

The value of the constant depends on the amount of gas in the sample and the temperature.

The Graph of Volume vs. Pressure in Figure 1A shows the type of curve obtained for a given quantity of gas at a fixed temperature. A linear relationship, Figure 1B, is obtained when the Volume is plotted versus $\frac{1}{\text{Pressure}}$. This linear relationship tells us that the volume is directly proportional to $\frac{1}{\text{Pressure}}$ or that volume is inversely proportional to pressure.

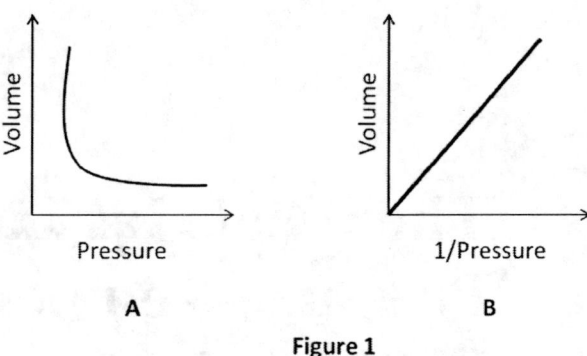

Figure 1

The Temperature-Volume Relationship

Charles's Law can be stated as follows: *The volume of a fixed amount of gas maintained at constant pressure is directly proportional to its absolute temperature.* Thus, doubling the absolute temperature, say from 250 K to 500 K, causes the gas volume to double. Since the relationship between temperature and pressure is directly proportional one would expect a linear plot of volume versus temperature Figure 2.

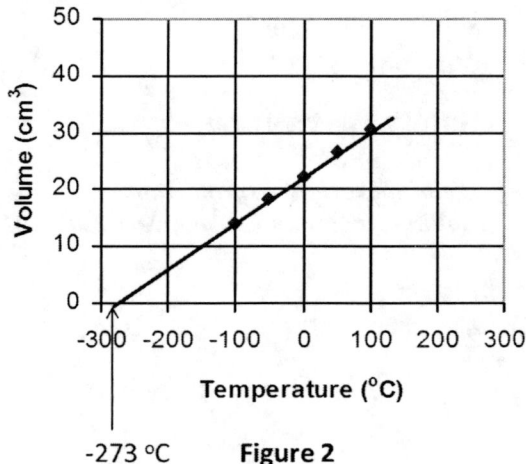

Figure 2

Expressing this relationship mathematically gives:

$$V = \text{constant} \times T \quad \text{or} \quad \frac{V}{T} = \text{constant} \qquad (2)$$

All gases change into the liquid or solid phase at low temperatures. If you extrapolate the line obtained from measuring the volume of the gas to the point at which the volume of the gas would have been zero, the temperature is –273 °C or absolute zero.

The Quantity-Volume Relationship

Avogadro's law states that *the volume of a gas maintained at constant temperature and pressure is directly proportional to the number of moles of the gas.* That is,

$$V = \text{constant} \times n \qquad (3)$$

Therefore, a linear relationship exists between the volume of a gas and its number of moles.

These three historically important gas laws were obtained by holding two variables constant in order to see how the other variables affect each other. Using the symbol $\propto$, which is read " is proportional to," we have

Boyle's law: $V \propto \dfrac{1}{P}$ (constant n, T)

Charles's law: $V \propto T$ (constant n, P)

Avogadro's law: $V \propto n$ (constant P, T)

The familiar Ideal gas law is obtained by first combining these relationships to make a more general gas law:

$$V \propto \frac{nT}{P}$$

If we call the proportionality constant R, we obtain

$$V = R\left(\frac{nT}{P}\right)$$

Rearranging, we have

PV = *n*RT (4)

In this experiment, we will study the relationship between the temperature of a gas sample and the pressure it exerts. Using the apparatus shown in Figure 3, we will place an Erlenmeyer flask containing an air sample in water baths of varying temperature. Pressure will be monitored with a pressure sensor and temperature will be monitored using a temperature probe. The volume of the gas sample and the number of moles it contains will be kept constant. Pressure and temperature data pairs will be collected during the experiment and then analyzed. From the data and graph, you will determine what kind of mathematical relationship exists between the pressure and absolute temperature of a confined gas. You will then use your data to find a value for absolute zero on the Celsius temperature scale.

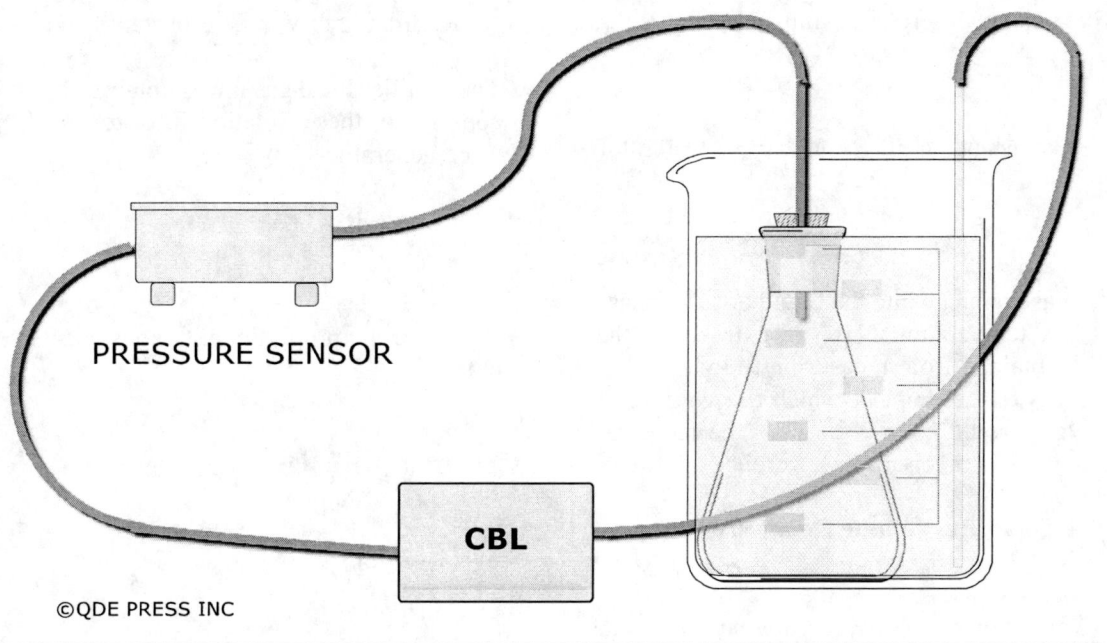

Figure 3

Procedure

1. Work in pairs
   - Wear goggles
   - Obtain from your TA a CBL system

2. Prepare a boiling-water bath. Put about 800 mL of hot tap water into a 1-L beaker and place it on a hot plate. Turn the hot plate to a high setting.

3. Prepare an ice-water bath. Put about 700 mL of cold tap water into a second 1-L beaker and add ice.

4. Put about 800 mL of room-temperature water into a third 1-L beaker.

5. Put about 800 mL of hot tap water into a fourth 1-L beaker.

6. Prepare the temperature probe and pressure sensor for data collection.

   - Plug the temperature probe into an adapter cable in Channel 1 of the CBL.
   - Plug the pressure sensor into an adapter cable in Channel 2 of the CBL. A 30-45 cm piece of heavy-wall plastic tubing is already connected to the end opening of the 3-way valve of the pressure sensor, as shown in Figure 3.
   - Open the side arm of the pressure sensor valve to allow air to enter and exit. Open the plastic valve of the pressure sensor by aligning the blue handle with the arm that leads to the pressure sensor (see Fig. 4).

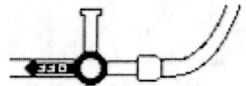

Figure 4

- Insert a 1-hole stopper fitted with a glass tube into a 125-mL flask. Twist the stopper to ensure a tight fit. Attach the plastic tubing to the glass tube in the stopper.
- Close the side arm of the pressure sensor valve by aligning the blue handle of the valve with the side arm, as shown in Figure 3. The air sample to be studied is now confined in the flask.

7. Turn on the CBL unit and the calculator. Start the CHEMBIO program and proceed to the MAIN MENU.

8. Set up the calculator and CBL for a temperature probe and calibration (in °C), and a pressure sensor and calibration (in atmospheres).

   - Select SET UP PROBES from the MAIN MENU.
   - Enter "2" as the number of probes.
   - Select TEMPERATURE from the SELECT PROBE menu.
   - Enter "1" as the channel number.
   - Select USE STORED from the CALIBRATION menu.
   - Select PRESSURE from the SELECT PROBE menu.
   - Enter "2" as the channel number.
   - Select USE STORED from the CALIBRATION menu.
   - Select ATM from the PRESSURE UNITS menu.

9. Set up the calculator and CBL for data collection.

   - Select COLLECT DATA from the MAIN MENU.
   - Select TRIGGER from the DATA COLLECTION menu.

10. Collect pressure vs. temperature data for your gas sample.

    - Place the flask into the ice-water bath. Make sure the entire flask is covered (see Figure 3). Stir.
    - Place the temperature probe into the ice-water bath.
    - Monitor pressure and temperature on the CBL screen by pressing the [CH VIEW] button on the CBL. When "CH1" in the upper-left corner of the CBL screen blinks, the *Channel 1* temperature (in °C) is displayed on the CBL. When you press [CH VIEW] again, "CH2" starts to blink—the *Channel 2* pressure (in atm) is now displayed on the CBL. Continue to press [CH VIEW] to alternate between the two readings.
    - When the temperature and pressure readings displayed on the CBL screen have both stabilized, press [TRIGGER] on the CBL to store the pressure-temperature data pair.

11. Select MORE DATA from the Trigger menu. Repeat the Step 10 procedure using the room-temperature bath.

12. Select MORE DATA from the Trigger menu. Repeat the Step 10 procedure using the hot-water bath.

13. Select MORE DATA from the Trigger menu. Use a ring stand, utility clamp, and slit stopper to suspend the temperature probe in the boiling-water bath. **CAUTION:** *Do not burn yourself or the probe wires with the hot plate.* To keep from burning your hand, hold the tubing of the flask using paper towels. After the temperature probe has been in the boiling water for a few seconds, place the flask into the boiling-water bath and repeat Step 10. Remove the flask and the temperature probe after you have pressed [TRIGGER] and finished collecting data. Select STOP from the Trigger menu and press [ENTER].

14. Examine the data points along the displayed graph of pressure vs. temperature (°C). As you move the cursor right or left, the temperature (X) and pressure (Y) values of each data point are displayed below the graph. Record the data pairs in your data table. Round pressure to the nearest 0.01 atm and temperature to the nearest 0.1°C.

15. Examine your graph of pressure vs. temperature (°C). In order to determine if the relationship between pressure and temperature is direct or inverse, you must use an absolute temperature scale; that is, a temperature scale whose 0° point corresponds to absolute zero. We will use the Kelvin absolute temperature scale. Instead of manually adding 273° to each of the Celsius temperatures to obtain Kelvin values, we will create a new list for Kelvin temperature, L3:

    - Press [ENTER] to return to the MAIN MENU.
    - Select QUIT to quit the data-collection program.
    - Create a list of Kelvin temperature values in L 3:

    TI-82 or TI-83 Calculators:

    To view the lists, press [STAT] to display the EDIT menu and then select Edit. Move the cursor up and to the right until the L3 heading is highlighted. Create a list of Kelvin temperature values in L 3 by pressing [2nd] [L1] + 273 [ENTER]. L1 is temperature (°C), L 2 is pressure (atm), and L 3 is temperature (K). Record the Kelvin temperature values in your data table.

16. Follow this procedure to calculate regression statistics and to plot a best-fit regression line on your graph of pressure vs. temperature (K):

    - Start the CHEMBIO program again and proceed to the MAIN MENU. **Important:** Do *not* select SET UP PROBES on the MAIN MENU—doing so will clear the data lists.
    - Select FIT CURVE from the MAIN MENU.
    - Select LINEAR L 3, L2. The linear-regression statistics for these two lists are displayed for the equation in the form:

    $$y = ax + b$$

    where x is temperature (K), y is pressure, a is a proportionality constant, and b is the y-intercept.

    - To display the power-regression curve on the graph of pressure vs. temperature (K), press [ENTER], then select SCALE FROM 0 from the SCALE DATA menu. If the relationship between pressure and absolute temperature is a direct relationship, the curve should be linear and pass through (or near) the origin. Examine your graph to see if this is true for your data.

The data that you have collected can also be used to confirm the value for absolute zero on the Celsius temperature scale. Instead of plotting pressure versus Kelvin temperature like we did above, this time you will plot Celsius temperature on the y-axis and pressure on the x-axis. Since absolute zero is the temperature at which the pressure theoretically becomes equal to zero, the temperature where the regression line (the extension of the temperature-pressure curve) intercepts the y-axis should be the Celsius temperature value for absolute zero. You can use the data you collected in this experiment to determine a value for absolute zero

- Press [ENTER] to return to the MAIN MENU and select QUIT.
- To plot a graph with temperature (°C) on the vertical axis and pressure on the horizontal axis, press [2nd] [STAT PLOT], then select 1:Plot1. Use the arrow keys to position the cursor on each of the following Plot1 settings. Press [ENTER] to select any of the settings you change: Plot1 = On, Type = ⋮⋮ , Xlist = L 2, Ylist = L 1, and Mark = ▫ .
- To plot a graph of temperature (°C) vs. pressure, press [ZOOM], then select ZoomStat.
- Press [STAT] [▶] to display the CALC menu. Select LinReg (ax+b).
- To identify the lists that correspond to the two variables, press [2nd] [L 2] [,] [2nd] [L1] [ENTER]. The statistics are displayed for the equation in the form:

$y = ax + b$

where x is pressure, y is temperature (°C), a is a proportionality constant, and b is the y-intercept. Note the value of the y-intercept (in °C).
- To display a best-fit regression line on the graph of pressure vs. temperature (°C), first press [Y=]. Press [CLEAR] to clear the Y1= equation, then press [VARS]. Select Statistics, then press [▶] [▶] to display the EQ menu. Select RegEQ to copy the linear regression equation to Y1=.
- Press [WINDOW] and rescale the pressure by entering 0 as Xmin and 1.5 as Xmax. Rescale the temperature by entering -300 as Ymin and 100 as Ymax. To enter -300, use [(-)], not [-].
- Press [GRAPH] to view the graph of pressure vs. temperature (°C) with a best-fit regression line.
- To interpolate along the regression line, press [TRACE], then press [▲] once. A cursor is displayed on the regression line, along with its X and Y coordinates below the graph. Use [◀] to move the cursor left along the regression line to a pressure (Y) value that is equal to 0 atm. The temperature (in °C) at this pressure is equivalent to absolute zero.
- Record your result for this Temperature in your data sheet.

Experiment 7　　　　　　　　　　　　　　Ideal Gas Law: Pressure-Temperature Relationship

Worksheet 7

Name _____

Date _____ Lab Instructor/Section _____

Pre-lab	____/20
Data	____/20
Post-lab	____/20
Safety/Part.	____/40
Total	____/100

Data sheet

Water bath	Pressure (atm)	Temperature (°C)	Temperature (K)	Constant (P/T or P×T)

(P/T or P×T)

Ice　　_____　　_____　　_____　　_____

Room　　_____　　_____　　_____　　_____

Hot　　_____　　_____　　_____　　_____

Boiling　　_____　　_____　　_____　　_____

Temperature, in degrees Celsius at which the pressure is 0 atm. _____

- 61 -

Experiment 7 — Ideal Gas Law: Pressure-Temperature Relationship

Questions

1. In order to perform this experiment, what two experimental factors were kept constant?

2. Based on the data and graph that you obtained for this experiment express in words the relationship between gas pressure and temperature.

3. Write an equation to express the relationship between pressure and temperature (K). Use the symbols P, T, and k.

4. One way to determine if a relationship is inverse or direct is to find a proportionality constant, k, from the data. If this relationship is direct, k = P/T. If it is inverse, k = P x T. Based on your answer to Question 3, choose one of these formulas and calculate *k* for the four ordered pairs in your data table (divide or multiply the P and T values). Show the answer in the fourth column of the Data and Calculations table. How "constant" were your values?

Experiment 7   Ideal Gas Law: Pressure-Temperature Relationship

Pre-Laboratory 7

Name _____

Date _____ Lab Instructor/Section _____

Pre-laboratory

1. Complete the following table for one mole of an ideal gas at a constant volume.

Pressure	Temperature
1.0 atm	273 K
2.5 atm	
3.8 atm	

2. From the information calculated above, derive the mathematical form of the **Temperature-Pressure Relationship and evaluate the constant.** (i.e. see for example the Temperature-Volume relationship in the Principles section)

Experiment 8

Molar Mass by Vapor Density

Objective

To measure the molar mass of an unknown liquid using the ideal gas law.

Equipment

Unknown liquid	250-mL Erlenmeyer flask	Thermometer
Aluminum foil	Large graduated cylinder	
1-L beaker	Utility clamp	

Safety Precautions

Wear approved eye protection. Keep the flask under the hood after immersing it in the hot water.

Principles

The ideal gas law provides a simple means of determining the molar mass of gaseous substances and liquids that can be vaporized at low temperatures. In this experiment, the molar mass of a volatile liquid will be determined. The liquid is converted into the gaseous state at an elevated temperature so it completely fills a vessel, the volume of which can be measured. Upon cooling the liquid condenses and may be weighed. From the measured pressure (P, in atm), temperature (T, in K) and volume (V, in L), of the gaseous sample, the number of moles of gas (n) may be calculated.

$$n = \frac{PV}{RT} \quad (1)$$

where the gas constant R is 0.0821 L atm/mol K. By combining the number of moles of gas calculated in this manner, with the weight of the sample (m), the molar mass (M, g/mol), of the compound may be found.

$$n = \frac{m}{M} \text{ or } M = \frac{m}{n} \quad (2)$$

Procedure

All masses in this experiment should be made with an accuracy of 0.01 g. First, heat about 750 mL of water in a 1 L beaker to near boiling. While the water heats, obtain a piece of aluminum foil from the side shelf. From the measuring bottles on the side shelf, place approximately 3 mL of one of the unknown samples into a clean, dry 250-mL Erlenmeyer flask. Cover the mouth of the flask tightly with the piece of aluminum foil. Make a small hole in the center of the foil with a pin or toothpick, see Figure 1. Under the hood, lower the flask into the hot water and clamp it firmly.

The water should surround the flask as completely as possible, but must not reach the lower edge of the foil. Boil the water gently and record the temperature of the water in the Data Sheet. Once all the liquid in the flask has vaporized, continue to heat for three more minutes. Then remove the flask from the bath, dry it, set it on a towel and allow it to cool to room temperature.

Carefully weigh (to 0.01 g) the foil-capped dry flask and its contents, i.e. the condensed liquid.

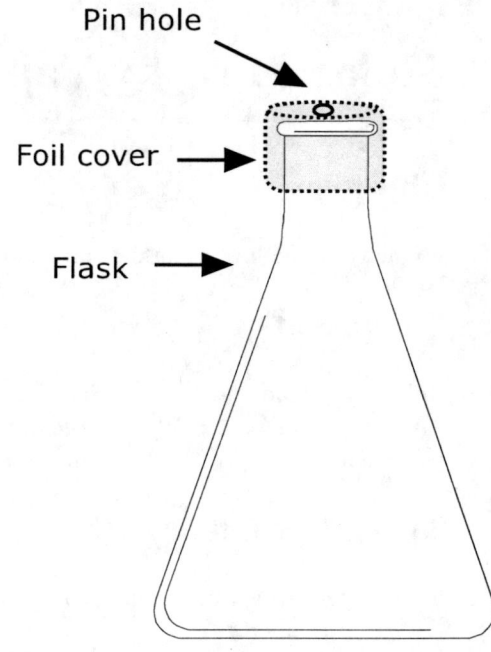

© QDE PRESS INC

Figure 1

The mass of this condensed liquid is the mass of the vapor that completely filled the flask at atmospheric pressure and at the boiling point of water. The unknown liquid should be poured out of the flask, the liquid wetting the flask allowed to evaporate, and the dry flask and foil weighed together. Be sure all of the unknown liquid has evaporated. Fill the flask with water. Then pour the water into a large graduated cylinder, thus determining the volume of the flask. This volume is also the volume of the vapor at experimental conditions.

Experiment 8　　　　　　　　　　　　　　　　　Molar Mass by Vapor Density

Worksheet 8

Name _____

Date _____ Lab Instructor/Section _____

Pre-lab	____/20
Data	____/20
Post-lab	20 /20
Safety/Part.	____/40
Total	____/100

Data sheet

1. Temperature of water
   °C   _____

2. Mass of flask, foil and condensed vapor
   g   _____

3. Mass of empty flask plus foil
   g   _____

4. Mass of condensed vapor
   g   _____

5. Volume of water required to fill the flask
   (that is equal to the volume of the gas)
   mL   _____

6. Barometric Pressure
   mm Hg   _____

7. Molar mass of compound (see Equations 1
   and 2 - remember to use the proper units for the
   ideal gas equation. Show your work.)   _____

Pre-Laboratory 8

Name _____

Date_____ Lab Instructor/Section_____

Pre-laboratory

The mass of a gas is determined as in the experiment described in Experiment 8. The mass is determined to be 2.01 g. The volume of the water, which is equal to the volume of the gas, is measured to be 243 mL, the boiling water temperature was 99.8°C and the barometric pressure was 755 mm Hg. Calculate the molar mass of the gas.

Experiment 9

Experimental Determination of R, the Ideal Gas Constant

Objective

Experimentally determine the value of the ideal gas constant, R. Learn how to measure the volume of a gas by water displacement and correct the pressure of the collected gas by subtraction of the partial pressure of water vapor.

Equipment

Florence flask
1-L beaker
1-L graduated cylinder
Potassium chlorate
Manganese dioxide
Test tube
Utility clamp
Bunsen burner
Custom flask stopper

Safety Precautions

Wear approved eye protection. Before heating the test tube containing $KClO_3$, make sure that the pinch clamp is OPEN – never heat a closed system.

Principles

The postulate made by Avogadro, that equal volumes of gases contain the same number of molecules, proved to be an important key to the evaluation of molar masses of gaseous compounds. From Avogadro's hypothesis, it follows that one mole of any gas will occupy the same volume when measured at the same temperature and pressure. Thus, the molar volume of oxygen, hydrogen, and chlorine or any other gaseous element or compound is the same, while the mass of one mole of each of these gases is different. By measuring the volume (V, in liters), mass (m, in g), pressure (P, in atm) and temperature (T, in K) of a gas of known molar mass (M, g/mol), the value of the ideal gas constant, R, is determined. Once R has been determined for a gas of known molar mass, the molar mass of an unknown gas can be determined by the same measurements, this time solving the below equation for M.

$$PV = nRT = \frac{m}{M}RT$$

$$R = \frac{MPV}{mT} \qquad (1)$$

Temperature °C	H₂O Vapor Pressure mm of Mercury	Temperature °C	H₂O Vapor Pressure mm of Mercury
10	9.2	20	17.5
11	9.8	21	18.7
12	10.5	22	19.8
13	11.0	23	21.1
14	12.0	24	22.4
15	12.8	25	23.8
16	13.6	26	25.2
17	14.5	27	26.8
18	15.5	28	28.4
19	16.5	29	30.1

Table 1

In this experiment the thermal decomposition of potassium chlorate (catalyzed by manganese dioxide) will be used to prepare pure oxygen.

$$2KClO_3(s) \xrightarrow{MnO_2} 2KCl(s) + 3O_2(g)$$

The difference in the mass of the solid before and after heating is equal to the mass (m) of oxygen gas liberated.

The oxygen will be collected over water and its volume (V) measured at a known temperature (T) and pressure. When gases are collected by displacement of water, part of the total pressure in the flask is due to the vapor pressure of water. The pressure of oxygen (P) will be calculated using the total pressure in the room the day the experiment is carried out (measure with a barometer) and the vapor pressure of water at the measured temperature. Dalton's law of partial pressures as applied to this case is

$$P_{total} = P_{O_2} + P_{H_2O}$$

The partial pressure of water, P_{H_2O}, is given in Table 1 at several temperatures.

Procedure

Work in pairs. From the side shelf, obtain a 1-liter Florence flask with a two-hole stopper filled with two pieces of bent glass tubing. To one piece of bent tubing (the one with the long tube inside the flask) will be attached a piece of rubber tubing with a pinch clamp. The rubber tubing will deliver water that is displaced from the flask and should be placed into a 1-liter beaker that you should get from the side shelf.

Fill the Florence flask two thirds with tap water, insert the stopper into the neck, and loosen the pinch clamp. In order to fill the delivery tube with water, one partner should turn the Florence flask on its side (shorter tubing up, delivery tubing down) so that water flows through the delivery tubing. When the delivery tubing is filled the other partner should tighten the pinch clamp.

Weigh out approximately 1.2 g of potassium chlorate (KClO₃) and place it in a small test tube. To the potassium chlorate add a small amount of manganese dioxide (MnO₂) and rotate the tube to mix the two solids. Weigh the test tube and contents accurately (to 0.01 g) and record the mass in the Data Section. Place the test tube in an extension clamp with the open end elevated about two inches above the bottom of the tube. Connect the test tube to the main apparatus by inserting the rubber stopper on the shorter bent tubing attached to the flask into the test tube, see Figure 1.

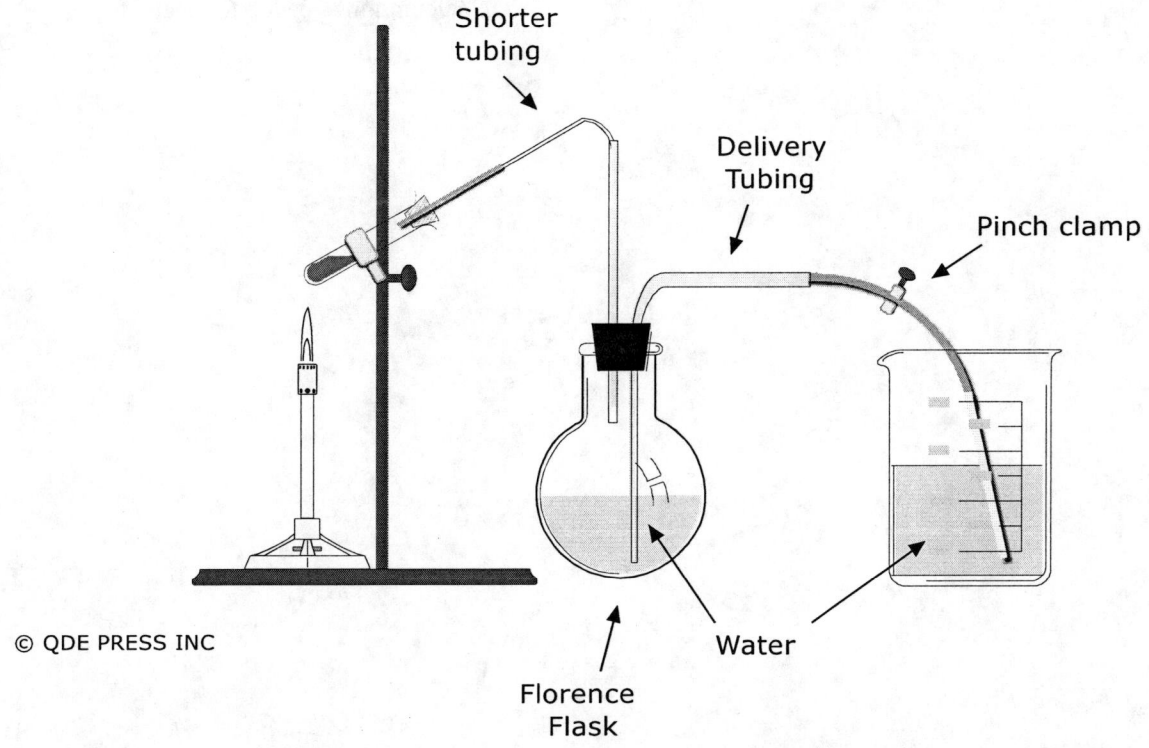

Figure 1

Next fill the beaker about half full of water and raise the beaker to equalize the top of the water level in it with the top of the water level in the Florence flask. With the water levels equal, open the pinch clamp momentarily and close it again. This operation makes the pressure in the flask the same as in the laboratory. *Discard the water in the beaker*, replace the beaker under the delivery tube and again open the pinch clamp. If only a small amount of water runs into the beaker, the system is leak-free and ready to use. If the water runs continuously, close the pinch clamp, stop the leaks and repeat the procedure outlined in this paragraph.

Check the assemblage of your apparatus with the Figure and have your instructor check it before heating is started.

Make sure the pinch clamp is open before heating the test tube. Heat the test tube slowly until water starts running into the beaker. Increase the heat when the flow decreases and continue heating until the flow finally stops. Stop heating at this point and allow the apparatus to cool to room temperature - be sure the delivery tube remains under the water level in the beaker during the cooling period – as the test tube cools, some water will flow back into the flask
($PV = nRT$).

When the test tube has cooled, again equalize the top of the water levels in the beaker and flask-closing the pinch clamp while the levels are equal. Pour the water in the beaker into a large graduated cylinder to determine its volume - this volume of water corresponds to the volume of oxygen produced in the chemical reaction. Record this measurement in the Data Sheet.

Detach the test tube with its contents from the rest of the apparatus and weigh the tube plus contents accurately (to 0.01 g). Record this mass in the Data Section. Obtain the atmospheric pressure and temperature from your instructor and make these entries. Perform the calculations needed to calculate R.

Experiment 9   Experimental Determination of R, The Ideal Gas Constant

Worksheet 9

Name _____

Date _____ Lab Instructor/Section _____

Pre-lab	____/20
Data	____/20
Post-lab	____/20
Safety/Part.	____/40
Total	____/100

Data Section

1. Initial mass of test tube + $KClO_3$ + MnO_2 _____ g

2. Mass of test tube and remaining residue after heating   _____ g

3. Mass of oxygen produced in the reaction   _____ g

4. Volume of water collected in the experiment,   _____ mL
   which equals the volume of oxygen produced in the reaction

5. Temperature   _____ K

6. Barometric pressure   _____ mm Hg

7. Partial pressure of O_2, correcting the barometric   _____ mm Hg
   pressure for the vapor pressure of the water

8. Pressure of O_2 in atmospheres   _____ atm

9. Value of R in the units L, atm, K, mol,   _____
   showing those units clearly with your calculated value

Calculations:

Don't forget the post-lab question on the back.

Question

1. Indicate whether the follow experimental mistakes will increase or decrease the calculated value of R; explain your answers.

(a) In adjusting the **final** water level after the O_2 was produced, the student adjusted the levels to the bottom of the water in the beaker and the top of the water in the flask (rather than correctly the top of each).

(b) A student completed the experiment without waiting for the test tube to cool after heating.

Experiment 9   Experimental Determination of R, The Ideal Gas Constant

Pre-Laboratory 9

Name _____

Date _____ Lab Instructor/Section _____

Pre-laboratory

1. In an experiment as described in the Principles section, 158 mL of O_2 is collected over water. The mass of the O_2 is measured to be 0.207 g, the temperature is 290 K and the barometric pressure that day is 754 mm Hg. Calculate the value of R in the units L, atm, K, mol, showing those units clearly with your calculated value. Remember to correct the pressure of O_2 for the vapor pressure of the water.

Experiment 10

Paper Chromatography

Objective

To learn the principles of paper chromatography and use the method to separate iron(III), copper(II) and nickel(II) in order to identify whether they are present in an unknown sample.

Equipment

Chromatography paper
Acetone
Dimethylglyoxime
1 M FeCl$_3$
1 M CuSO$_4$
1 M NiSO$_4$
Ammonia
1.0 M HCl
Solution of Ni^{2+}, Fe^{3+} and Cu^{2+}
Pencil
Ruler

Safety Precautions

Wear approved eye protection. Avoid inhaling chemicals; work under the hood with the ammonia solution.

Principles

The usual first step in the chemical analysis of an unknown sample is the separation of each of the components that are present in the unknown. Chromatography, a word derived from the Greek *chroma* meaning color, is a powerful separation technique that has been used in different forms by chemists for nearly one hundred years. It works on the principle that different substances adsorb differently on the same inert material. The mixture to be separated is generally dissolved in a solvent that carries it past the inert material. Because of differential adsorption, as the mixture is carried along the inert material the components of the mixture separate. The components that adsorb to the chromatography paper to a lesser degree will move the fastest while the components that adsorb the strongest move slowest. The solvent can be either a liquid (as in paper and column chromatography) or a gas (as in vapor phase chromatography) and the inert material can be either a solid or a high boiling liquid. For a given solvent system used on the same adsorbent at a constant temperature, the amount any given substance moves relative to the leading edge of the solvent will be characteristic of that substance. The movement of a substance is expressed as a constant *retention factor*, R$_F$, value where

$$R_F = \frac{\text{Distance component moves}}{\text{Distance solvent moves}}$$

In this experiment, iron(III), copper(II) and nickel(II) will be separated by paper chromatography and identified in an unknown by comparison of both R_F values and colors. The Fe^{3+} will produce rust colors, the Cu^{2+} will form a blue complex when treated with ammonia and Ni^{2+} will form a red complex when treated with dimethylglyoxime. Known samples of each of the ions will be run as well as a mixture of all three and an unknown sample that could contain any mixture of the three. The behavior of the metal ions in the known samples will be used as a basis in identifying the components in the unknown sample.

Procedure

Under a hood, carefully mix together 24 mL of acetone and 10 mL of dilute hydrochloric acid. **CAUTION** - *acetone is extremely flammable, HCl fumes are harmful and the liquid burns the skin - if you get any on your skin, wash immediately with copious quantities of water*. Pour this solvent system into a 1-L beaker and cover the beaker tightly with a plastic film.

Obtain a piece of chromatography paper measuring 28 cm by 12 cm. Draw a line in pencil (**not pen, a pencil must be used**) about one cm away from a long edge of the paper. This line will indicate the starting point for the samples.

Starting on the left and leaving about 3 cm of paper at each end, place small equally spaced drops of each of the five samples listed below on the penciled line, Figure 1. Use a capillary tube to apply each spot, *using a new capillary tube for each sample*. Note that it is important to use only a small drop of each solution. Make sure that each drop is placed on the line. With a pencil, circle each spot directly on the filter paper. The samples are:

a. A sample containing a soluble Fe^{3+} compound.

b. A sample containing a soluble Cu^{2+} compound.

c. A sample containing a soluble Ni^{2+} compound.

d. A sample containing a mixture of all three ions.

e. Unknown sample on side shelf that may contain one, two or all three of the ions.

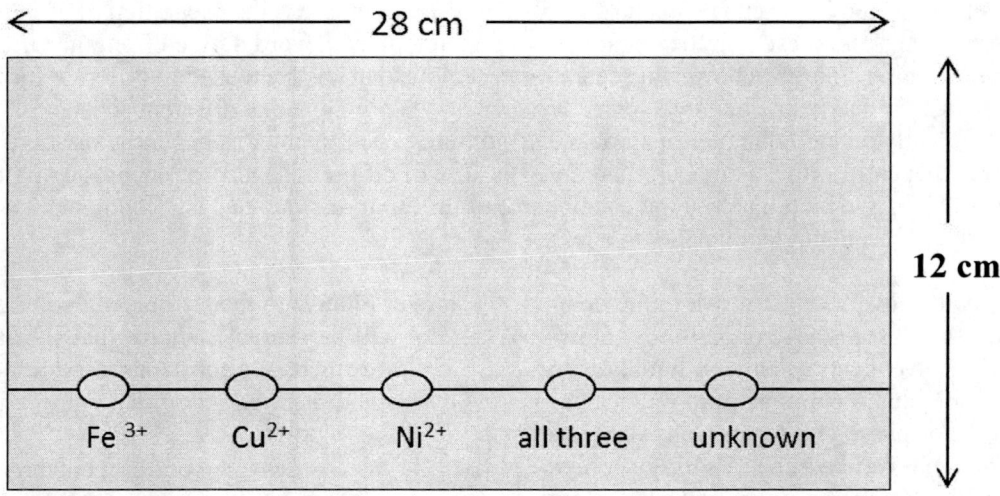

Figure 1

After all the spots have dried, fold the paper accordion-style making sure no spots are on folds and place it carefully in the beaker. The penciled line should be above the solvent mixture. Replace the plastic film cover over the beaker and wait until the solvent has moved up the paper about 8 cm from the penciled line. Remove the paper quickly, marking the position of the leading edge of the solvent with a pencil, and allow the solvent to evaporate completely under a hood. Measure the distance from the *starting line* to the leading edge of the solvent and record that number on the data sheet.

Circle any spots that appear at this time. Under the fume hood carefully pour a few milliliters of concentrated ammonia (sometimes labeled NH_4OH) into a 400-mL beaker and hold the chromatogram over the dish for one full minute. Circle any new spots that become visible on contact with the gaseous ammonia. *While still moist with ammonia* spray the paper quickly with dimethylglyoxime. If Ni^{2+} does not show red, hold the chromatogram over the NH_4OH again while still wet with the dimethylglyoxime solution. Dry the paper and circle any new spots. Do this carefully, holding the paper as high as possible in the hood. Record on the data sheet the color, which reagent, if any, made each of the spot appear, and the distance each spot is from the starting line. Use the approximate center of your circle around each spot to determine this distance. Calculate the R_F value (show work) for each of the spots and record these values with the appropriate metal ion on the Data Sheet. For samples d and e you should be able to identify each spot by its color, the time in the development procedure that the spot appeared and the R_F value.

Clearly indicate which ions were present in sample e, your unknown. After completely filling in the data sheet, *staple your chromatogram to your data sheet*, and turn them in to your instructor.

Experiment 10   Paper Chromatography

Worksheet 10

Name _____

Date _____ Lab Instructor/Section _____

Pre-lab	____/20
Data	____/20
Post-lab	20 /20
Safety/Part.	____/40
Total	____/100

Data sheet

Unknown # _____

Distance from pencil line to the leading edge of the solvent: _____

		Color of spot	Which reagent (if any) made the color appear	Distance the center of the spot is from the starting line	R_F value
Spot a.	Fe^{3+}				
Spot b.	Cu^{2+}				
Spot c.	Ni^{2+}				
Spot d.	Fe^{3+}				
	Cu^{2+}				
	Ni^{2+}				
Spot e.	Fe^{3+}				
	Cu^{2+}				
	Ni^{2+}				

Metal ions in the unknown sample _____

R_F **Calculations:**

85

Experiment 10 Paper Chromatography

Pre-Laboratory 10

Name _____

Date _____ Lab Instructor/Section _____

Pre-laboratory

Use Figure 2 shown below to calculate the R_F values (back of this page) for compounds x, y, and z. Measure the distances with the provider ruler, recording the distance the solvent front moved and how far each "spot" moved. Show your calculations. Using the Figure to show an example, the R_F for ion x is = d_x/D_s. Write your answers on the back.

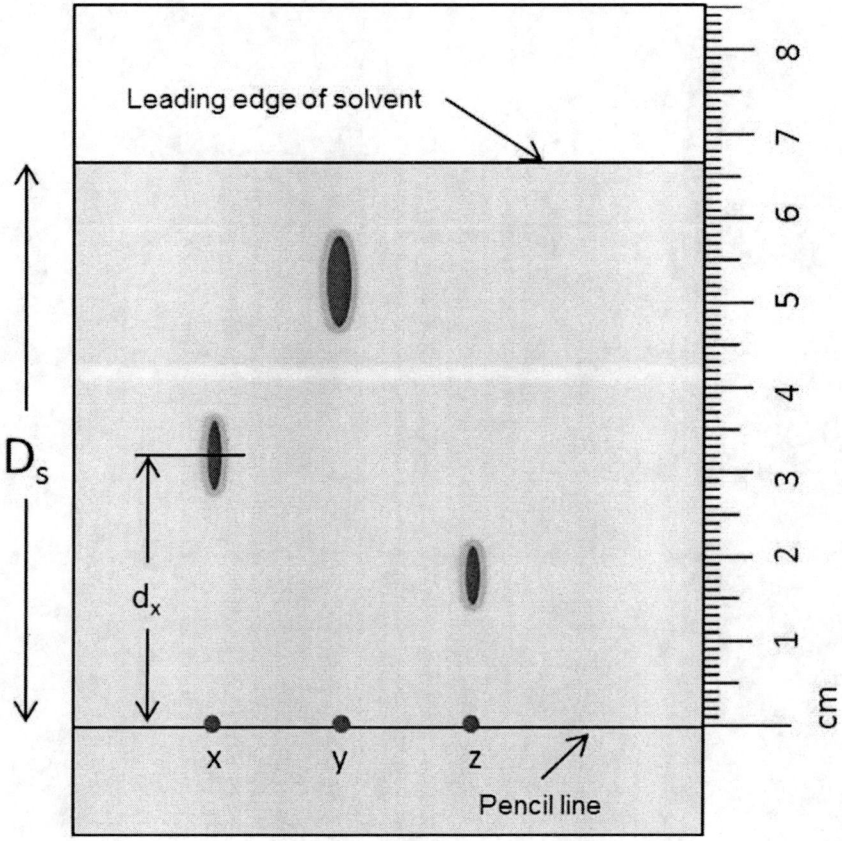

Figure 2

Experiment 10　　　　　　　　　　　　　　　　　　　　　　　　Paper Chromatography

Distance solvent front moved (D_s) _____

Distance ion x moved (d_x)　　　　　_____　　　$R_F(x) =$ _____

Distance ion y moved　　　　　　　_____　　　$R_F(y) =$ _____

Distance ion z moved　　　　　　　_____　　　$R_F(z) =$ _____

Experiment 11

Determination of Waters of Hydration

Objective

To determine the mass percent water and number of water molecules in a hydrated ionic compound.

Equipment and Chemicals

Unknowns
Crucibles with covers
Triangle holder
Bunsen burner
Tongs

Safety Precautions

Wear approved eye protection. The crucible gets very hot, do not burn your hands.

Principles

Many ionic compounds unite chemically with water, ammonia and other polar molecules. Those that unite with water are called hydrates. For example, zinc sulfate combines with water to form crystalline $ZnSO_4 \cdot 7 H_2O$. This compound is stable under normal atmospheric conditions. All pure samples of this hydrate show the same percentage of water. On heating, a sample of a hydrate may lose all its water of hydration, reverting to the anhydrous compound (the original compound with no waters of hydration).

In this experiment, a hydrate of unknown composition will be heated to remove the waters of hydration and both the percentage of water and the number (n) of waters of hydration will be determined. In these calculations, the loss of mass upon heating (the water or steam released) is divided by the mass of the sample which, when multiplied by 100%, is the percentage of water in the compound. Then the number of moles of water given off may be calculated by dividing the loss of mass by 18 g/mol (molar mass of water). The molar mass of your anhydrous compound is given in this experiment. With this value, the number of moles of the anhydrous compound may be determined by dividing the mass of the anhydrous compound by its molar mass.

Dividing the number of moles of water by the number of moles of anhydrous compound gives the number (n) of waters of hydration.

The molar mass of the anhydrous salt of your unknown hydrate will be one of the following:

A --- 258 g/mol   C --- 161 g/mol

B --- 208 g/mol   D --- 120 g/mol

Procedure

Work in pairs. Clean and dry a porcelain crucible and cover. Place the empty, covered crucible on a triangle holder and heat it until a cherry redness appears. Allow the crucible to cool to room temperature and determine its mass to the nearest 0.01 of a gram. As the first crucible is cooling, repeat the procedure with a second crucible. Obtain a sample of an unknown hydrate from your instructor. Add about 1 to 1.5 grams of the hydrate to the crucible and determine the mass of the covered crucible with the sample to the same accuracy as before. Record the two masses in the Data Section.

Place the covered crucible with the sample on the triangle and heat gently for a few minutes. Start with gentle heating so that there will be no loss of material from spattering during the initial heating. Continue to heat for about 15 minutes with the hottest part of the burner flame. Allow the covered crucible to cool until it has reached room temperature. Now determine the mass of the covered crucible and residue. Reheat the crucible for 5 minutes, cool and reweigh. Repeat the heating, cooling and reweighing until two consecutive weighings are the same within 0.01 grams.

As the first crucible with sample is cooling, repeat the entire process with a sample of a different hydrate. This second trial may be run simultaneously with the first trial to save time. When one crucible is cooling another may be heated or weighed.

Experiment 11 Determination of Waters of Hydration

Worksheet 11

Name _____

Date _____ Lab Instructor/Section _____

Pre-lab	____/20
Data	____/20
Post-lab	20 /20
Safety/Part.	____/40
Total	____/100

Data sheet

	Sample Letter ____	Sample Letter ____
1. Mass of crucible and cover	_____ g	_____ g
2. Mass of crucible, cover and sample	_____ g	_____ g
3. Mass of hydrated compound = 2−1	_____ g	_____ g
4. Mass of crucible, cover and sample after heating	_____ g	_____ g
5. Mass of anhydrous compound = (4−1)	_____ g	_____ g
6. Mass of water = (3 − 5)	_____ g	_____ g
7. Percentage of water = (6/3 × 100%)	_____ %	_____ %
8. Moles of water = (6/molar mass H_2O)	_____ mol	_____ mol
9. Molar mass of anhydrous compound	_____ g/mol	_____ g/mol
10. Moles of anhydrous compound = (5/9)	_____ mol	_____ mol
11. Waters of hydration (n) = (8/10)	_____	_____

- 91 -

Experiment 11 — Determination of Waters of Hydration

Pre-Laboratory 11

Name _____

Date _____ Lab Instructor/Section _____

Pre-laboratory

A 1.55 g sample of a hydrated $ZnSO_4 \cdot nH_2O$ compound is heated in a crucible to remove the water and after the heating weighs 1.08 g. Calculate the percent water in the hydrated sample, the moles of water, the moles of anhydrous $ZnSO_4$ (molar mass = 161.4 g/mol) produced in the reaction and the value of n, the number of water molecules in the hydrated sample. Remember that n is a whole number.

Percent water _____   Moles water _____

Moles $ZnSO_4$ _____   Value of n _____

Experiment 12

Shapes of Molecules

Objective

Use the VSEPR model to predict the shapes of molecules. To use these shapes to assign the hybridization of central atoms and the overall polarity of the molecules.

Equipment

Completed at home.

Principles

The arrangement in space of the atoms in a molecule is an important component of its chemical bonding description. For example, CH_4 could be a flat planar molecule (incorrect) or tetrahedral (correct) as pictured below (the wedges mean that these H atoms are above the plane of the paper, the dash lines mean they are below with the carbon atom and hydrogen atoms connected by solid lines in the plane of the paper).

```
        H                              H
        |                              |
   H — C — H                      H — C ....H
        |                             /
        H                            H
      planar                      tetrahedral
```

The chemical and physical properties of CH_4 and its many derivatives will be influenced greatly by the shape.

A simple model, based on Lewis structures, is extremely useful in predicting the shapes of many molecules; the valence-shell electron-pair repulsion (VSEPR) model. The main premise of the model is based on the idea that the electron pairs about an atom repel each other. The VSEPR model predicts the shape around each **central atom**, an atom in a
molecule that is bonded to at least two other atoms. Starting with the Lewis structure, count the number of **lone pairs** on the central atom plus the number of **atoms bonded to it** (*not bonds*). This sum determines the **bonded-atom lone-pair arrangement** (also called the **electron pair arrangement**), the shape that maximizes the distances between the valence-shell electron pairs. For example, a central atom with two bonded atoms and no lone pairs will be linear because this shape puts the areas of electron density in the bonds as far apart as possible. Other shapes for electron-pair arrangements greater than two are in the textbook. One important point needs to be emphasized before beginning the experiment.

The VSEPR model predicts shapes based on the number of bonded atoms and lone pairs of electrons around a central atom; the bonded-atom lone-pair arrangement. The actual **molecular shape** is frequently different because the lone pairs, although influencing the geometry, are not part of the final molecular shape. Molecular shape is always described in terms of the position of the atoms only. For example, the central nitrogen atom in NH_3 has three bonded atoms and one lone pair; the bonded-atom lone-pair arrangement is a tetrahedral. The molecular shape, as pictured, is a trigonal pyramid because the area occupied by the lone pair is not part of the molecular shape. In contrast, BF_3 has three bonded atoms and no lone pairs, so both its bonded-atom lone-pair arrangement and molecular shape are trigonal planar. The bonded-atom lone-pair arrangement decides the actual molecular shape, but is different from it if lone pairs are present on the central atom.

Once the bonded-atom lone-pair arrangement has been determined, the hybridization of atomic orbitals on that central atom is also known. For example, if the bonded-atom lone-pair arrangement is linear, the hybridization is *sp*; consult your text for additional arrangements. Another important question that can be answered once the actual molecular shape is known is whether the molecule is polar or nonpolar. Polarity has a great influence on the chemical properties.

Procedure

<u>Part I</u>

Draw a correct Lewis structure, count the number of valence electrons, count the number of atoms bonded to the central atom and the lone pairs on the central atom, determine the steric number and assign the correct bonded-atom lone-pair arrangement to the first compound, NF_3, listed in the tables below. Determine the bond angles in the molecule and name the molecular shape (i.e., the ammonia pictured in the introduction was a trigonal (3-sided) pyramid). Assign the hybridization of the central atom, the number of sigma and pi bonds and determine if the molecule is polar or nonpolar. Determine the bond order for the bonds listed in the appropriate box. Carry out this procedure for all of the molecules or ions listed in Part I. Part II and the Pre-Laboratory exercise should also be completed.

Experiment 12　　　　　　　　　　　　　　　　　　　　　　Shapes of Molecules

Worksheet 12

Name _____

Date _____ Lab Instructor/Section _____   Total ____/100

Data sheet

Part I

Molecule A NF₃	Lewis structure	Hybridization
Number of valence electrons		Number of σ and π bonds σ:　　π:
Number of bonded atoms on central atom		Molecular shape
Number of lone pairs on central atom	Bonded-atom lone-pair arrangement	Polarity
Central atom steric number	Bond angles	Bond order N-F:

Molecule B HCCl₃	Lewis structure	Hybridization
Number of valence electrons		Number of σ and π bonds σ:　　π:
Number of bonded atoms on central atom		Molecular shape
Number of lone pairs on central atom	Bonded-atom lone-pair arrangement	Polarity
Central atom steric number	Bond angles	Bond order C-H:　　C-Cl:

- 97 -

Molecule C **H₂CO**	Lewis structure	Hybridization
Number of valence electrons		Number of σ and π bonds σ: π:
Number of bonded atoms on central atom		Molecular shape
Number of lone pairs on central atom	Bonded-atom lone-pair arrangement	Polarity
Central atom steric number	Bond angles	Bond order C-H: C-O:

Ion D **[NO₂]⁻**	Lewis structure	Hybridization
Number of valence electrons		Number of σ and π bonds σ: π:
Number of bonded atoms on central atom		Molecular shape
Number of lone pairs on central atom	Bonded-atom lone-pair arrangement	Polarity
Central atom steric number	Bond angles	Bond order N-O:

Experiment 12 — Shapes of Molecules

Ion E [H₃O]⁺	Lewis structure	Hybridization
Number of valence electrons		Number of σ and π bonds σ: π:
Number of bonded atoms on central atom		Molecular shape
Number of lone pairs on central atom	Bonded-atom lone-pair arrangement	Polarity
Central atom steric number	Bond angles	Bond order O-H:

Molecule F BrNS	Lewis structure	Hybridization
Number of valence electrons		Number of σ and π bonds σ: π:
Number of bonded atoms on central atom		Molecular shape
Number of lone pairs on central atom	Bonded-atom lone-pair arrangement	Polarity
Central atom steric number	Bond angles	Bond order N-Br: N-S:

Molecule G **PF₅**	Lewis structure	Hybridization
Number of valence electrons		Number of σ and π bonds σ: π:
Number of bonded atoms on central atom		Molecular shape
Number of lone pairs on central atom	Bonded-atom lone-pair arrangement	Polarity
Central atom steric number	Bond angles	Bond order P-F:

Molecule H **TeF₆**	Lewis structure	Hybridization
Number of valence electrons		Number of σ and π bonds σ: π:
Number of bonded atoms on central atom		Molecular shape
Number of lone pairs on central atom	Bonded-atom lone-pair arrangement	Polarity
Central atom steric number	Bond angles	Bond order Te-F:

Worksheet 12

Name _____

Date _____ Lab Instructor/Section _____

Data sheet
Part II

Draw the Lewis structures and assign the correct molecular shapes of the following molecules and ions.

H_2O $[NO_3]^-$ CO_2
SO_2 $[NH_4]^+$ $[BrF_4]^+$
SF_6 HCN $[ICl_4]^-$

Experiment 12　　　　　　　　　　　　　　　　　　　　　　　　Shapes of Molecules

Pre-Laboratory 12

Name _____

Date _____ Lab Instructor/Section _____

Pre-laboratory

1. Draw the Lewis structures for H_2O and HCN.

2. Fill in the table below.

Compound	Number of atoms bonded to central atom	Number of lone pairs on central atom	Bonded-atom lone-pair arrangement	Molecular shape	Hybridization	Polarity
H_2O						
HCN						

Chemistry 112L

Lab Experiments

SAFETY POLICIES

Safe practice in the chemical laboratory is a mutual responsibility and requires the full cooperation of everyone concerned at all times. This cooperation means that each student and instructor will observe safety precautions and procedures. The following general safety rules will be rigidly and impartially enforced throughout the semester. Noncompliance may result in dismissal from the lab and/or may result in a grading penalty.

1. Appropriate safety glasses must be worn at all times anywhere in the laboratory, even when not performing an experiment. Contact lenses should not be worn during the lab period.

2. Footwear should provide adequate protection against possible safety hazards (broken glass, reagent spills, etc.)

3. Food or drink will not be allowed in the laboratory.

4. Horseplay or other acts of carelessness are prohibited.

5. Unauthorized experiments are not permitted. Unapproved variations in experiments, including changes in the quantities of reagents, may be dangerous.

6. Every student is responsible for keeping his work area neat and orderly. After the experiment is over, clean the equipment and store it away correctly before you leave.

7. The instructor should be informed immediately of any safety hazards or accidents.

All accidents have causes and therefore can be prevented. Pay careful attention to what you are doing in each experiment, follow all instructions and use common sense. Be aware of what your neighbors do, you may be a victim of their accidents. Do not hesitate to comment tactfully to a neighbor whom you observe engaging in an unsafe practice. Thoroughly acquaint your-self with the location and use of emergency equipment (fire extinguishers, eye-wash stations, showers, etc.) around the lab. With the positive approach of good safety practice, all personal injuries can be avoided.

I have read and understood the safety rules outlined above. I agree to abide by them at all times while participation in Chemistry _____ laboratory, Section _____.

(Signed)_____ (Date)_____

Teaching Assistants Name_____

Safety and Laboratory Techniques

EMERGENCY & HEALTH INFORMATION

Please print.

Your Name _____

Campus Phone_____

Campus or Home address_____

_____
_____

Person to be contacted in case of an accident:

Name:_____

Phone:_____

Do you have any health problems or disabilities that may cause difficulties in the Chemistry laboratory? If so, please describe them briefly below. This information will remain confidential.

_____
_____
_____
_____
_____
_____

Contact lenses in Chemical Laboratories

The following is taken form "Handbook of Laboratory Safety" N.V. Stoek, Ed, 2nd ed., Chemical Rubber Co., Cleveland OH, 1971:

"Contact lenses worn by persons working in laboratories can increase injury from chemical splashes because the wearer may not be able to remove the lenses to permit thorough irrigation, and a person giving first aid may not know that contact lenses are being worn or how to remove them. It is recommend that contact lenses not be worn in laboratories in which chemicals are handled or that wearers be sure to use full eye protection at all times."

In the pamphlet, "Use of Contact Lenses in Industry," published by the Council on Occupational Health of the American Medical Association, there are three paragraphs, which are particularly applicable to wearing contact lenses in laboratories.

"Many physicians believe that the substitution of contact lenses for spectacles in industrial workers is contraindicated in workers whose eyes may be exposed to dusts, molten metals, or irritant chemicals. Small foreign bodies, which normally are washed away by tears, sometimes become lodged beneath contact lenses where they may cause injury to the cornea. Similarly chemicals splashed into the eye may be trapped under a contact lens and cause extensive corneal damage before the lens can be removed and the eye adequately irrigated."

"For effective protection for the eyes, the contact lens wearer should use in addition to his contact lenses the same approved face shields, conventional safety spectacles, or goggles for protection against job hazards, as would any other worker on a similar job. Since removal of a contact lens for urgent irrigation after injury is made is so difficult by spasms of the eyelids, the contact lens wearer is in even greater need of these protections than his or her counterpart who does not wear contact lenses, if the job carries high potential risk of eye injury."

"Contact lenses are not in themselves protective devices in fact may increase the degree of injury to the eyes. Contact lens wearers in similar employment should wear the same eye-protective device used by other workers."

I have been informed of and understood the hazards associated with wearing contact lenses in the laboratory. I agree to wear safety goggles at all times while participating in Chemistry _____ laboratory.

Signed _____ Date _____

Experiment 13

Qualitative Inorganic Analysis

Objective

To identify ions that are present in unknown solutions and solids using "wet chemical" separation methods. These methods are based on the behavior of different ions when they are allowed to react with certain reagents. *Reagents* are substances chosen because of their chemical activity with the ions being analyzed. Learning the chemistry that governs the identifications is an important part of this experiment.

Principles

Classical inorganic qualitative analysis has as its goal the identification of the species present in an unknown solution or solid. The unknowns contain ions whose identities need to be determined. In the classical analytical scheme the known properties of the different ions, both positive ions (cations) and negative ions (anions), are used to separate a mixture of them into successively smaller groups of ions, until some characteristic reaction may be used to confirm the presence or absence of each specific ion. In addition to analyzing the unknown for its component ions, the qualitative analysis scheme highlights some of the important chemical behavior of these metal salts in aqueous solution. The principles of chemical equilibrium are emphasized, as illustrated by precipitation reactions, acid-base reactions, complex-ion formation, and oxidation-reduction reactions. Students usually find the challenge of unknowns in qualitative analysis to be stimulating and (if possible) fun. Each experiment presents a puzzle that is solved "detective fashion" by assembling a collection of chemical clues into an airtight case for the correct identifications. As a bonus, the clues often take the form of colorful solutions and precipitates.

The formation and color of precipitates or the color of solutions are used to identify the ions. The principal separation technique is the stepwise precipitation of insoluble compounds, mainly insoluble hydroxides. Selective precipitation is followed by determining the relative tendencies of the solids to re-dissolve with either additional hydroxide ion or ammonia. The scheme was developed to illustrate the solution chemistry of the more common metals. All of the separations in this scheme involve systems in equilibrium and none are kinetically controlled (i.e., depend on the relative rates of two or more reactions).

The qualitative analytical scheme is divided into four parts:

1. Separation and identification of cations in a known solution.
2. Separation and identification of cations in a unknown solution.
3. Separation and identification of anions in known and unknown solutions.
4. Identification of both a cation and anion in an unknown salt.

Schedule—This experiment will last for five weeks.

Week	Samples Analyzed
1	Cation Known
2	Cation Unknown
3	Anion Known and Unknowns
4	Salt
5	Lab 13 Make-up Week

Recording Observations. In Parts 1 and 2, Cation Analysis, all observations and conclusions should be recorded in the Laboratory Manual on the blank spaces on each flow diagram. Record information for both your known samples and unknown samples, using the results for the known samples as a guide to identify the unknowns. In Part 3, Anion Analysis, fill in the spaces reserved in Table 1. An accurate, thoughtful record is important. Since the directions are brief you are encouraged to use your record as a guide to devise tests for completion of reactions. Understanding the equations representing the chemical reactions taking place and a little thought can save the student from errors.

Laboratory Reports. When you have identified the ions present in each of your unknown samples to your satisfaction, report it on the form provided at the end of this experiment.

General Techniques. The qualitative analysis scheme that follows is carried out on what is called a semimicro (small) scale of operations. This requires minimal quantities of reagents and sample, and the time necessary to carry out the operations is much less than that required using a macroscale (large) set of operations. The indications of reaction, appearance of a precipitate, development of color, etc., are all readily discernible on this semimicro scale. The usual reactor vessel on this scale is the 10 x 75 mm test tube. The quantities of solution handled are of the order of one to two milliliters, and the volumes of reagents to be added are on the order of one to several drops (~ 0.05 mL). Convenient reagent containers to use in this laboratory experiment are located on the side shelf.

1. **Separation of precipitates from solutions**—The principal separation method used is precipitation. This separation is performed using a centrifuge, which spins the sample in a small test tube at high speeds, causing the precipitate to settle rapidly to the bottom of the tube. In using the centrifuge always counterbalance the sample tube with a test tube containing water filled to the same level as the sample. Always place the cover on the centrifuge before turning it on (if a tube breaks, those nearby won't be hit by flying glass). *Allow the centrifuge to coast to a stop*. Do not attempt to slow it with your fingers. With most precipitates 30 - 60 seconds of centrifuging will be sufficient to obtain a good separation. If the solution still appears cloudy after this length of time, centrifuge for another minute or two. After centrifuging, most of the liquid may be poured into another container without disturbing the solid. This process is called decantation.

2. **Washing a precipitate**—Since decantation does not completely remove the solution it is recommended that the precipitate be washed following the separation. Washing is done by placing about 1 mL of water (or a recommended washing solution) into the tube containing the precipitate. Stir with a vertical motion with a stirring rod to suspend the solid in the washing solution. Centrifuge and decant the liquid.

3. **Addition of reagents**—The reagents are measured in drops delivered from a dropper bottle. When adding reagents the tip of the dropper must never touch the test solution or the walls of the test tube. A contaminated reagent solution can cause erroneous results.

4. **Heating and Cooling Solutions**—At several places in the procedure it is necessary to heat samples. *Never heat the test tube directly with a burner flame.* Directly heating the test tube will invari
ably result in the entire sample "bumping" out of the test tube. The proper way to heat a sample is to place the test tube into a small beaker of water and gently heating the beaker. Cooling a sample is done in the same manner, using cold water or ice in place of the hot water.

5. **Test for Completeness of Precipitation**—After adding the specified amount of a precipitating reagent it is suggested that a test for completeness of precipitation be carried out as follows. Centrifuge and without decanting the solution add a drop of reagent so it runs down the wall of the test tube. If precipitation is complete, no new precipitate will form when the reagent dissolves in the solution.

6. **Re-precipitation**—In some cases, particularly when the precipitate (ppt.) is gelatinous in nature, it may have to be dissolved and re-precipitated to separate salts that may have been carried down with the gelatinous precipitate. The resulting solution, after centrifuging (centr.) and decantation, is added to the solution resulting from the first precipitation or, discarded.

Part 1. Cation Analysis

Cations Studied—All present as the nitrate salts, 0.05 M

Ag^+, Al^{3+}, Fe^{3+}, Mg^{2+}, Ba^{2+}, Cu^{2+}, Ni^{2+}.

Reagents and Equipment—The reagents called for in this scheme for separations and for confirmatory tests are listed below:

Reagents

For Separations	For Confirmatory Tests
HCl (dil., 6 M)	KSCN (0.2 M)
HNO_3 (dil., 6 M)	Aluminon reagent (Note 1)
H_2SO_4 (dil., 6 M)	Na_2HPO_4 (0.2 M)
NH_3 (dil, 7.5 M)	Dimethylglyoxime (Note 2)
NaOH (dil., 6 M)	NH_4NO_3 (5 M)
H_2O_2 (3 %)	
KI (2 M)	
Na_2SO_3 (saturated)	

Note 1. Aluminon reagent (0.1 %) is prepared by dissolving 1 gram of ammonium aurintricarboxylate in a liter of distilled water.

Note 2. Dimethylglyoxime reagent (1 %) is prepared by dissolving 10 grams of dimethylglyoxime in 1 liter of 95 % ethyl alcohol. Fresh portions of hydrogen peroxide (H_2O_2)-and sodium sulfite (Na_2SO_3) should be obtained at the beginning of a laboratory period since on storage the peroxide tends to decompose and the sulfite may be oxidized to sulfate by dissolved oxygen.

General reference :

D. L. Reger, S. R. Goode and D. W. Ball, "Chemistry Principles and Practice", Third Edition, Brooks/Cole, Cengage Learning (Belmont, CA) 2010.

Basis of the Qualitative Analytical Separations in this Scheme--The separations used in this qualitative analytical scheme are based on the facts contained in an abbreviated set of solubility rules (see Reger, Goode and Ball, p 144) and the facts that some insoluble metal hydroxides dissolve in the presence of excess hydroxide ion (are amphoteric) and that some transition metal ions tend to form amine complexes with ammonia.

Brief Solubility Rules

1. All nitrates are soluble.
2. All chlorides are soluble except AgCl, Hg_2Cl_2 and $PbCl_2$.
3. All sulfates are soluble except $SrSO_4$, $BaSO_4$, Hg_2SO_4, and $PbSO_4$.
4. All carbonates are insoluble except those of the IA elements and NH_4^+.
5. All hydroxides are insoluble except those of the IA elements and NH_4^+, $Sr(OH)_2$ and $Ba(OH)_2$. $Ca(OH)_2$ is slightly insoluble.

*Insoluble compounds here are defined as those which precipitate upon mixing equal volumes of solutions 0.1 M in the corresponding ions.

The seven cations included for analysis in this scheme are Ag^+, Al^{3+}, Fe^{3+}, Mg^{2+}, Ba^{2+}, Cu^{2+}, and Ni^{2+}, listed roughly in the order in which they are separated. Aluminum hydroxide is amphoteric; its hydroxide reacts with bases as well as acids. The reaction is:

$$Al(OH)_{3\,(s)} + OH^- \rightleftarrows Al(OH)_4^-$$

The two cations in the group that tend to form complexes with ammonia under the conditions encountered in the laboratory are Cu^{2+} and Ni^{2+}. The reactions are:

$$Cu^{2+} + 4NH_3 \rightleftarrows Cu(NH_3)_4^{\,2+}$$

$$Ni^{2+} + 6NH_3 \rightleftarrows Ni(NH_3)_6^{\,2+}$$

Using these equilibria and the solubility rules, the seven cations will be separated into four groups

Group 1 HCl is added to the soluble nitrate salts of the metal ions (rule 1) to precipitate AgCl (rule 2).

Group 2 Addition of NH_3 makes the solution basic and precipitates the hydroxides of Al^{3+} and Fe^{3+} (rule 5). The excess NH_3 forms the soluble amine complexes of Cu^{2+} and Ni^{2+} keeping them in solution. A portion of the added NH_3 reacts with the HCl from the first separation to form NH_4^+ (ammonium cation), forming a buffer solution with the unreacted NH_3. This buffer prevents the hydroxide ion concentration from becoming large enough to precipitate the insoluble hydroxide $Mg(OH)_2$.

Group 3 This group precipitates in two steps. $BaSO_4$ is precipitated by adding sulfuric acid (rule 3 of the solubility table). The ammonia concentration in the solution containing the remaining cations, Mg^{2+}, Cu^{2+} and Ni^{2+} is increased to convert the last two cations into their amine complexes, and NaOH is added to furnish a large enough hydroxide ion concentration that $Mg(OH)_2$ now precipitates

Group 4 The precipitation of $Mg(OH)_2$ leaves only the Group 4 cations in solution as $Cu(NH_3)_4^{2+}$ and $Ni(NH_3)_6^{2+}$.

The foregoing separations of the cations into groups, the separations within the groups and the specific tests for identification of the individual ions are described in more detail on the page after each Group flow diagram. Refer to these more detailed description of the chemistry each step along the flow charts.

Characteristic Colors--Some of the transition metal compounds and complexes encountered here exhibit color. Attention to the colors (or their absence) of solutions and precipitates can often be quite helpful in deciding whether or not a particular ion might be present. The student should become familiar with the colors of the solutions and precipitates of the ions involved in the scheme. To this end, the colors of the ions, complexes, and precipitates encountered in this scheme are furnished below. Those ions that are colorless and those precipitates that are white are omitted from the following table.

Ion or Precipitate	Color
Fe^{3+} .	yellow to orange
$Fe(OH)_3$ (s)	red-brown
$FeSCN^{2+}$	red
Cu^{2+}	blue
$Cu(NH_3)_4^{2+}$	deep blue
Ni^{2+}	green – blue/green
$Ni(NH_3)_6$	violet

Flow Charts--The qualitative analytical scheme that follows is presented in the form of a flow chart. The student starts with a mixture of cations, as the soluble nitrates, page 117, all appearing in a rectangular box. Along a vertical line, just below the box, is the reagent to be added to the solution and its amount (2 d HCl). The 2 d HCl means that the student is to add *two* drops of *dilute* hydrochloric acid to the mixture in the box, with stirring. The vertical line in the diagram continues down to intersect with a horizontal line. This intersection means: centrifuge the solution after the above treatment, decant the supernatant solution into another test tube (indicated as a box to the right on the diagram, in this case solution 2A). The solid (Group) that is possibly precipitated by the reagent is indicated by an oval to the left on the diagram.

Both the known and unknown solutions will be analyzed by Groups. The experiment in each Group is initiated by analysis of a known solution containing all of the cations. The analysis of the unknown solutions using the following scheme sometimes leads to observations of very small quantities of precipitates and/or faint positive color tests for cations. Quite often these "trace" amounts are just that: small amounts of impurities or precipitates resulting from interfering cations. In deciding whether the test for
your unknown is positive or not, the student should compare the quantity of precipitate (or intensity of color) obtained with the unknown with that obtained in the analysis of a known solution; these should be comparable.

Procedure

Safety Precautions

Wear approved eye protection. Avoid spilling the test solutions on your hands or clothes

Part 1

Obtain a known test solution containing all of the cations from you're the side shelf. All of the cations are in the form of soluble nitrate compounds. Follow the procedures on the flow charts for separating and identifying the cations. Note your observations on appropriate locations on the chart. Use these observations when you analyze your unknown. Be sure to refer to the page following each flow chart in order to understand the chemistry involved in the tests.

Part 2

Obtain a general unknown sample that could contain metals in any of the Groups, and use the procedures outlined on the flow charts to determine the ions present. The general unknown contains three to five of the metal cations in it. Record the ions present on the answer sheet

Experiment 13   Qualitative Inorganic Analysis

Below is a **broad outline** of the separation of the seven cations into groups:

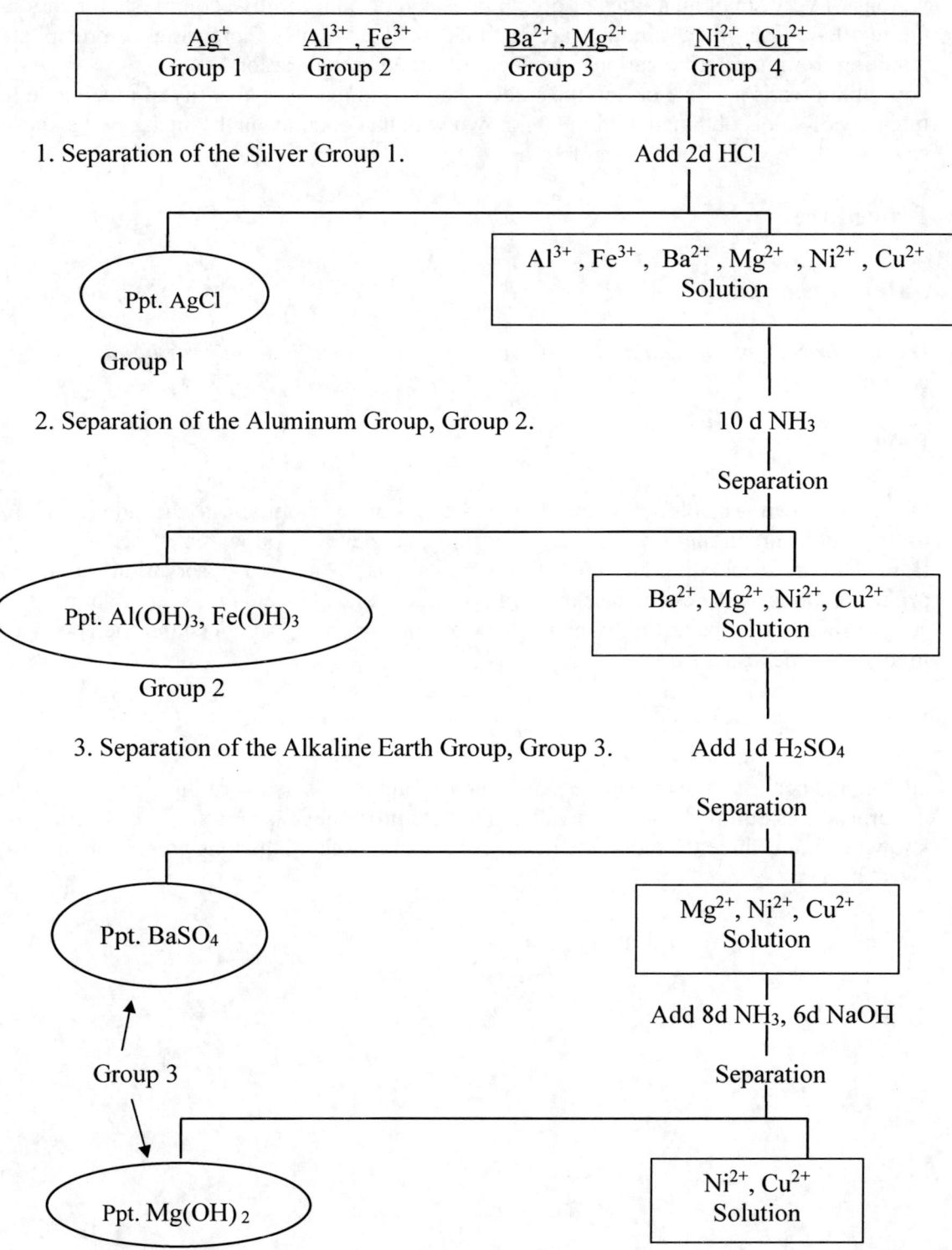

More specific directions for the separations of the ions within each group are given in the following pages.

Group 1- Precipitation and Identification of Silver

Start with 1 mL of the known or the unknown solution. Start all unknowns at the beginning.

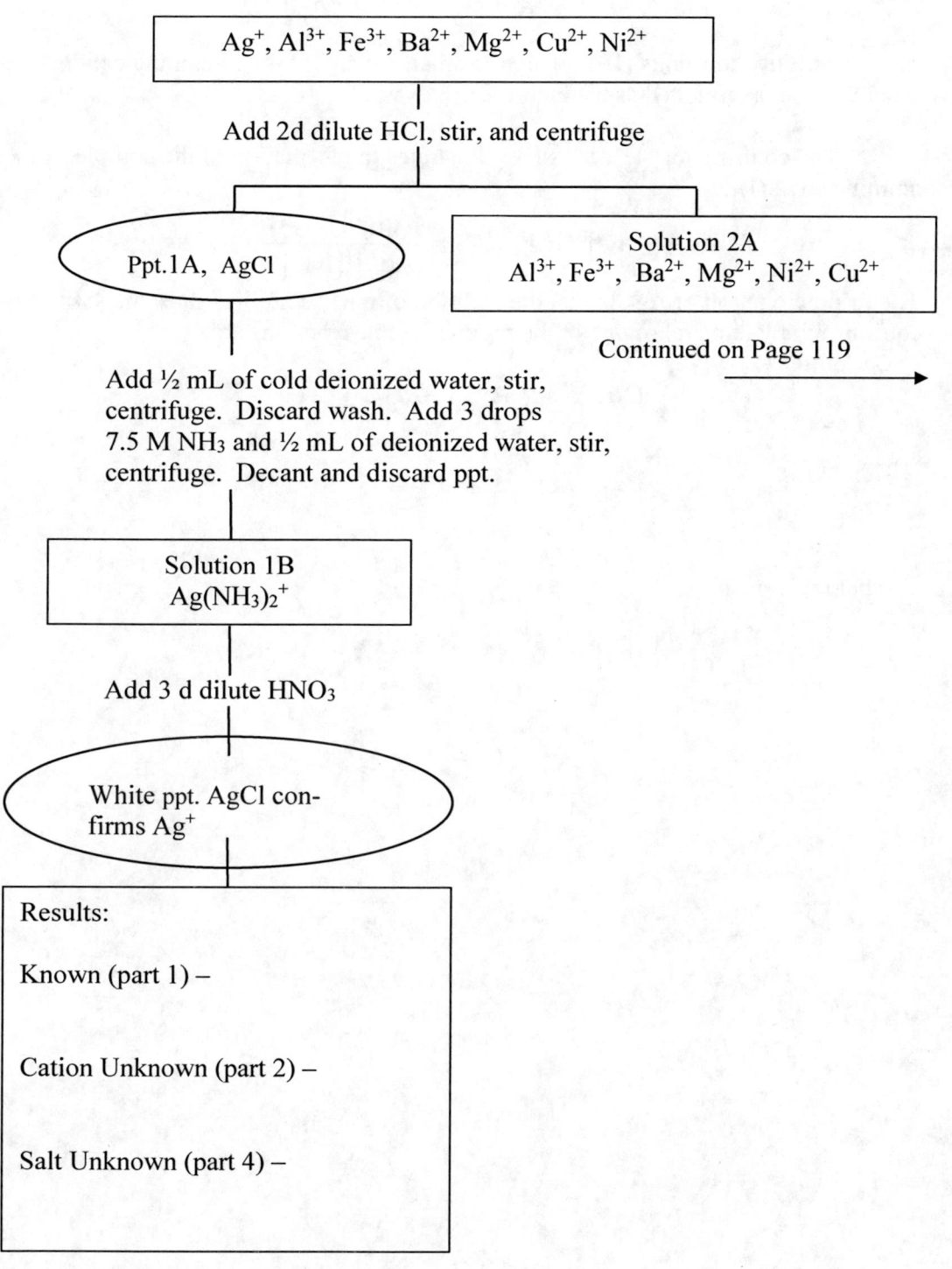

Results:

Known (part 1) –

Cation Unknown (part 2) –

Salt Unknown (part 4) –

Chemical Equilibria and Reactions Involved in the Cation Scheme.

Group 1 - Separation and Identification of Silver (Ag^+).

The reaction that occurs when HCl is added to the solution is:

$$Ag^+ + Cl^- \rightleftharpoons AgCl(s) \qquad K_1 = 1/K_{sp} = 5.6 \times 10^9$$

The insolubility constants (1/Ksp) are too small for any of the remaining cations in the scheme to be precipitated as the chlorides.

The confirmatory test for silver illustrates the formation of the complex cation, di-ammine silver(I)

$$Ag^+ + 2NH_3 \rightleftharpoons Ag(NH_3)_2^+ \qquad K = \frac{[Ag(NH_3)_2^+]}{[Ag^+][NH_3]^2} = 1.7 \times 10^7$$

The re-precipitation of AgCl upon the addition of nitric acid illustrates the shift of an equilibrium by removal of one of the products. The reactions are:

$$Ag(NH_3)^{2+} \rightleftharpoons Ag^+ + 2NH_3$$

$$2NH_3 + 2H^+ \rightleftharpoons 2NH_4^+$$

$$Ag^+ + Cl^- \rightleftharpoons AgCl(s)$$

The net reaction is:

$$Ag(NH_3)^{2+} + 2H^+ + Cl^- \rightleftharpoons AgCl + 2NH_4^+$$

Experiment 13   Qualitative Inorganic Analysis

Group 2 - Separation of Identification of Aluminum and Iron.

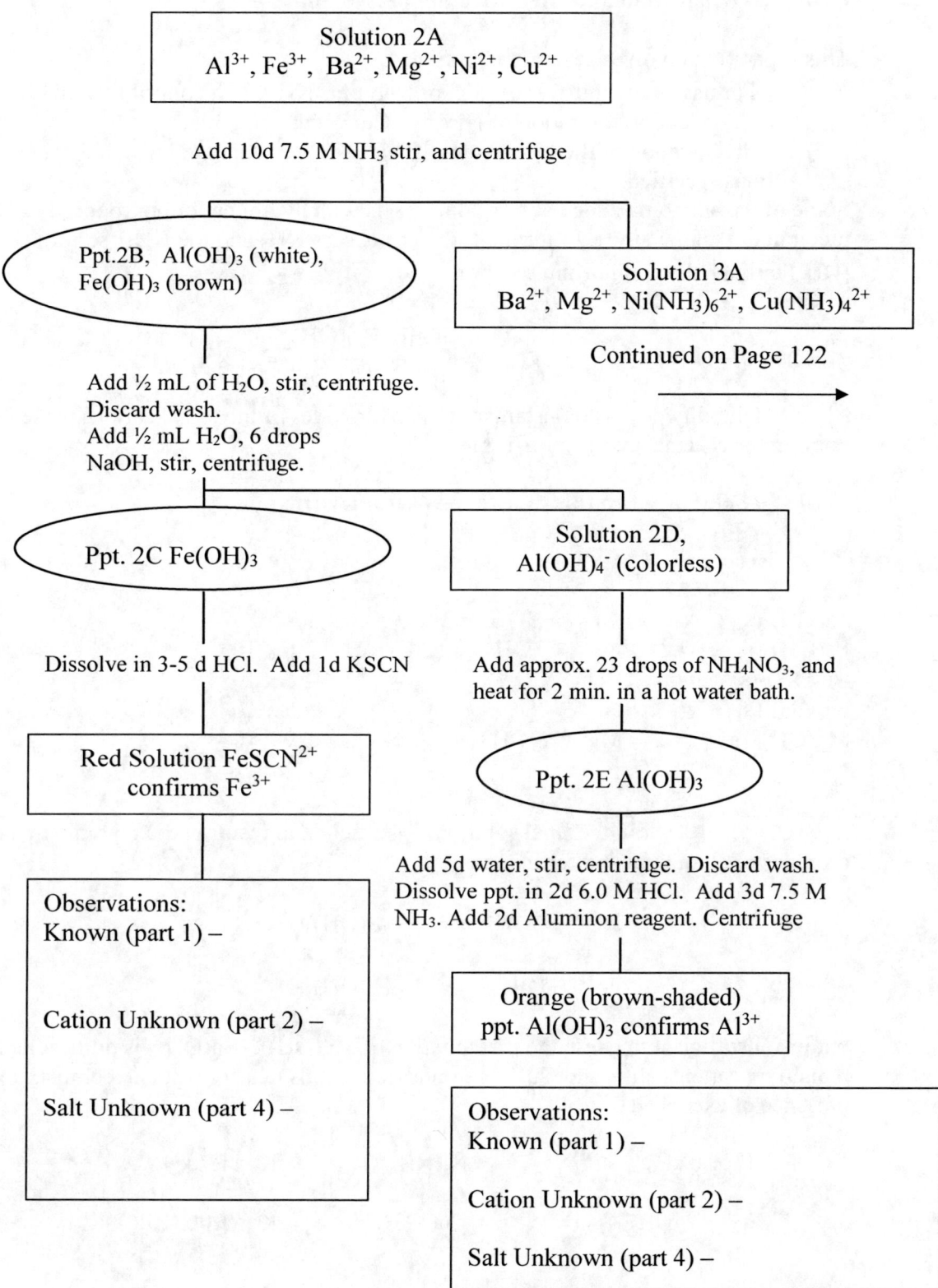

- 119 -

Chemical Equilibria and Reactions Involved in the Cation Scheme.

Group 2 - Separation and Identification of Al^{3+} and Fe^{3+}

This separation involves several factors:
- Formation of a buffer solution of NH_3 plus NH_4Cl to control the $[OH^-]$.
- Complex ion formation of Ni^{2+} and Cu^{2+} with ammonia.
- Precipitation of the hydroxides of Al^{3+} and Fe^{3+} that are insoluble under the conditions specified.

Some of the ammonia added at this point reacts with hydrogen ion, introduced as HCl in the precipitation of the silver, to form NH_4^+. The excess NH_3 and the NH_4^+ so formed governs $[OH^-]$ through the equilibrium reactions

$$NH_3 + H_2O \rightleftharpoons NH_4^+ + OH^- \qquad K = \frac{[NH_4^+][OH^-]}{[NH_3]} = 1.8 \times 10^{-5}$$

The solubility product constants for the hydroxides of the metal cations present at this point in the scheme are given in Table 1.

Table 1. Solubility Products for Some Metal Hydroxides.

Reactions			Ksp
$Al(OH)_3(s)$	$\rightleftharpoons$	$Al^{3+} + 3OH^-$	5×10^{-33}
$Fe(OH)_3(s)$	$\rightleftharpoons$	$Fe^{3+} + 3OH^-$	2.6×10^{-39}
$Cu(OH)_2(s)$	$\rightleftharpoons$	$Cu^{2+} + 2OH^-$	1.6×10^{-19}
$Ni(OH)_2(s)$	$\rightleftharpoons$	$Ni^{2+} + 2OH^-$	5.5×10^{-16}
$Mg(OH)_2(s)$	$\rightleftharpoons$	$Mg^{2+} + 2OH^-$	8.9×10^{-12}

The $[OH^-]$ (~ 10^{-5} M in the final solution) is sufficient to result in the precipitation of the hydroxides of aluminum and ferric ion,

$$Al^{3+} + 3OH^- \rightleftharpoons Al(OH)_3(s)$$

$$Fe^{3+} + 3OH^- \rightleftharpoons Fe(OH)_3(s)$$

but just insufficient to cause the precipitation of $Mg(OH)_2$. $Ba(OH)_2$ is quite soluble. The remaining cations, Ni^{2+}, and Cu^{2+}, are transition metals that form stable complex ions in the presence of excess NH_3.

$$Ni^{2+} + 6NH_3 \rightleftharpoons Ni(NH_3)_6^{2+} \qquad K = 1.7 \times 10^8$$

$$Cu^{2+} + 4NH_3 \rightleftharpoons Cu(NH_3)_4^{2+} \qquad K = 1.0 \times 10^{12}$$

Observe that ferric ion does not form an ammine complex that is sufficiently stable to prevent precipitation of the hydroxide. Although Cu^{2+} and Ni^{2+} form hydroxides, they do not precipitate from this solution because the formation of the ammine complexes reduces the concentration of the "uncomplexed" metal ions to a very low value. The copper-ammonia complex results in a deep blue solution and the nickel-ammonia complex is violet in solution.

Returning to the precipitate containing Al^{3+} and Fe^{3+}, as the hydroxides, treatment with excess sodium hydroxide dissolves $Al(OH)_3$ but not $Fe(OH)_3$. Al^{3+} is amphoteric, that is, it tends to form a stable anionic complex with hydroxide ion.

$$Al(OH)_3\,(s) + OH^- \rightleftarrows Al(OH)_4^-$$

The treatment with base results in the separation of ferric ion from aluminum, which remains in solution. The $Fe(OH)_3$ ppt. dissolves readily in hydrochloric acid. The confirmatory test for Fe^{3+}, treatment with potassium thiocyanate, results in formation of the deep red thiocyanatoiron (III) complex.

$$Fe^{3+} + SCN^- \rightleftarrows Fe(SCN)^{2+}$$

Addition of NH_4NO_3 to the basic solution containing $Al(OH)_4^-$ results in a significant reduction in the $[OH^-]$ due to the reaction

$$NH_4^+ + OH^- \rightleftarrows NH_3 + H_2O$$

This reaction breaks up the $Al(OH)_4^-$ complex and $Al(OH)_3$ precipitates.

This $Al(OH)_3$ precipitate is then re-dissolved in HCl, Aluminon, the ammonium salt of aurin tricarboxylic acid., forms an insoluble red chelate, or lake, with aluminum ion at pH 4 to 8. The precipitate, not of constant composition, is thought to be a mixture of aluminum hydroxide and the chelate itself.

3. Separation of the Group 3 Alkaline Earth Metal Cations From the Copper Group, Group 4

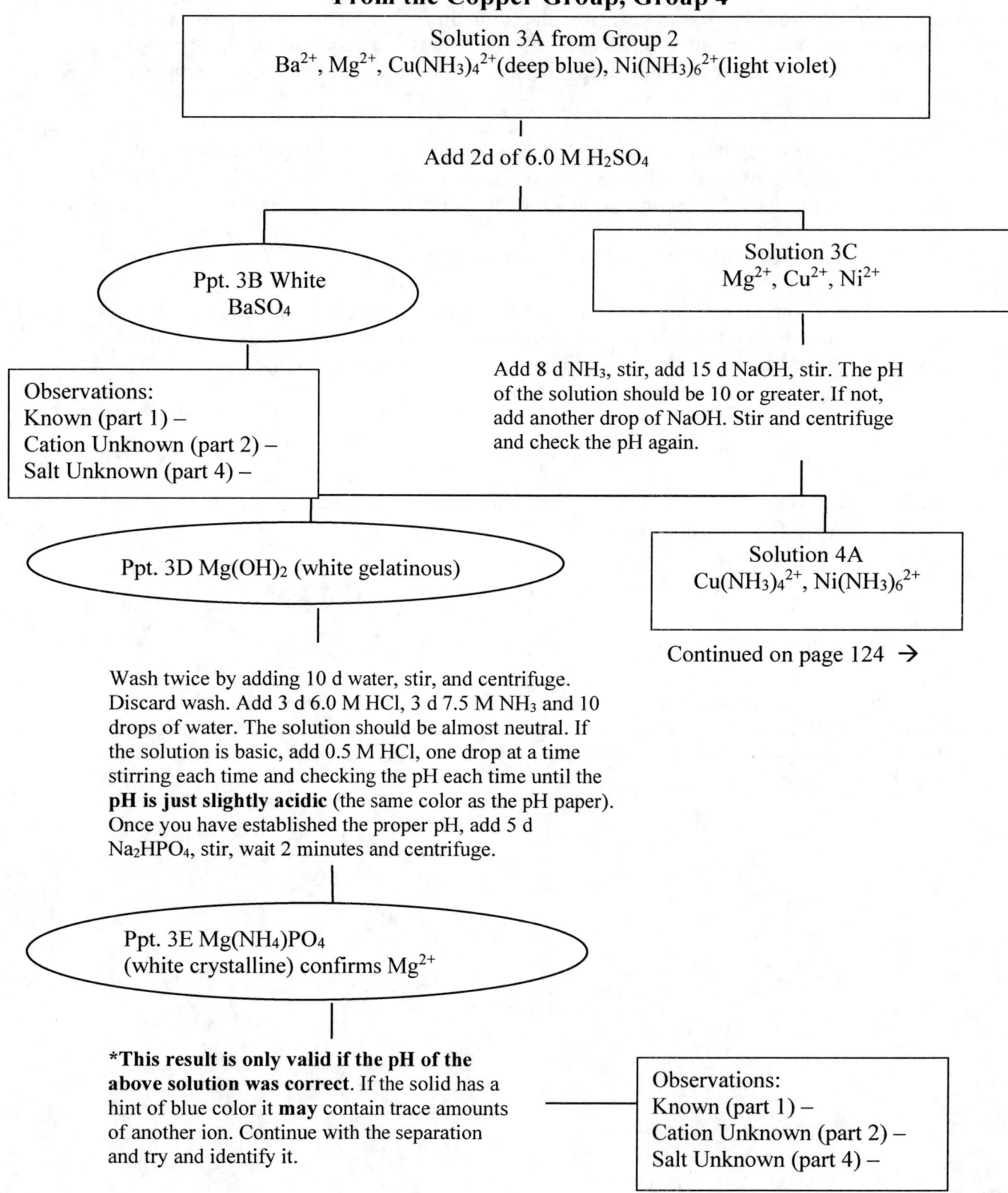

Chemical Equilibria and Reactions Involved in the Cation Scheme.

Group 3. Separation of Barium and Magnesium.

The addition of sulfuric acid to the solution remaining from the separation of the aluminum group results in precipitation of any barium present as white barium sulfate.

$$Ba^{2+} + SO_4^{2-} \rightleftarrows BaSO_4 (s), \quad K = 7 \times 10^8$$

A white precipitate at this point confirms the presence of Ba^{2+}. The acid also decomposes the transition metal ion-ammonia complexes that are present.

$$Cu(NH_3)_4^{2+} + 4H^+ \rightleftarrows Cu^{2+} + 4NH_4^+$$

This reaction can be viewed as occurring in two steps: (a) the dissociation of the copper ammine complex,

$$(a) \; Cu(NH_3)_4^{2+} \rightleftarrows Cu^{2+} + 4NH_3$$

and (b) the protonation of the free ammonia so formed, greatly lowering its concentration in solution.

$$(b) \; NH_3 + H^+ \rightleftarrows NH_4^+$$

The solution (containing Mg^{2+}, Cu^{2+} and Ni^{2+}) is now treated with more NH_3, reforming the ammine-complexes of copper and nickel. The solution, containing ammonium sulfate and ammonia, has a $[OH^-] \sim 10^{-5}$ M, so magnesium hydroxide does not precipitate, until the solution is made decidedly basic with the addition of sodium hydroxide:

$$Mg^{2+} + 2OH^- \rightleftarrows Mg(OH)_2 (s)$$

The gelatinous white $Mg(OH)_2$ will be contaminated with some occluded copper, and nickel. This occlusion will cause the precipitate to have a blue color if copper is present. Treatment of this precipitate with HCl dissolves it. Buffering is achieved with the addition of ammonia. Addition of the Na_2HPO_4 solution results in precipitation of white, finely divided $Mg(NH_4)PO_4$, confirmatory for magnesium. This process may require a minute or two to be completed.

$$NH_3 + HPO_4^{2-} \rightleftarrows NH_4^+ + PO_4^{3-}$$

$$Mg^{2+} + NH_4^+ + PO_4^{3-} \rightleftarrows Mg(NH_4)PO_4(s)$$

4. Separation and Identification of the Group 4 Cations

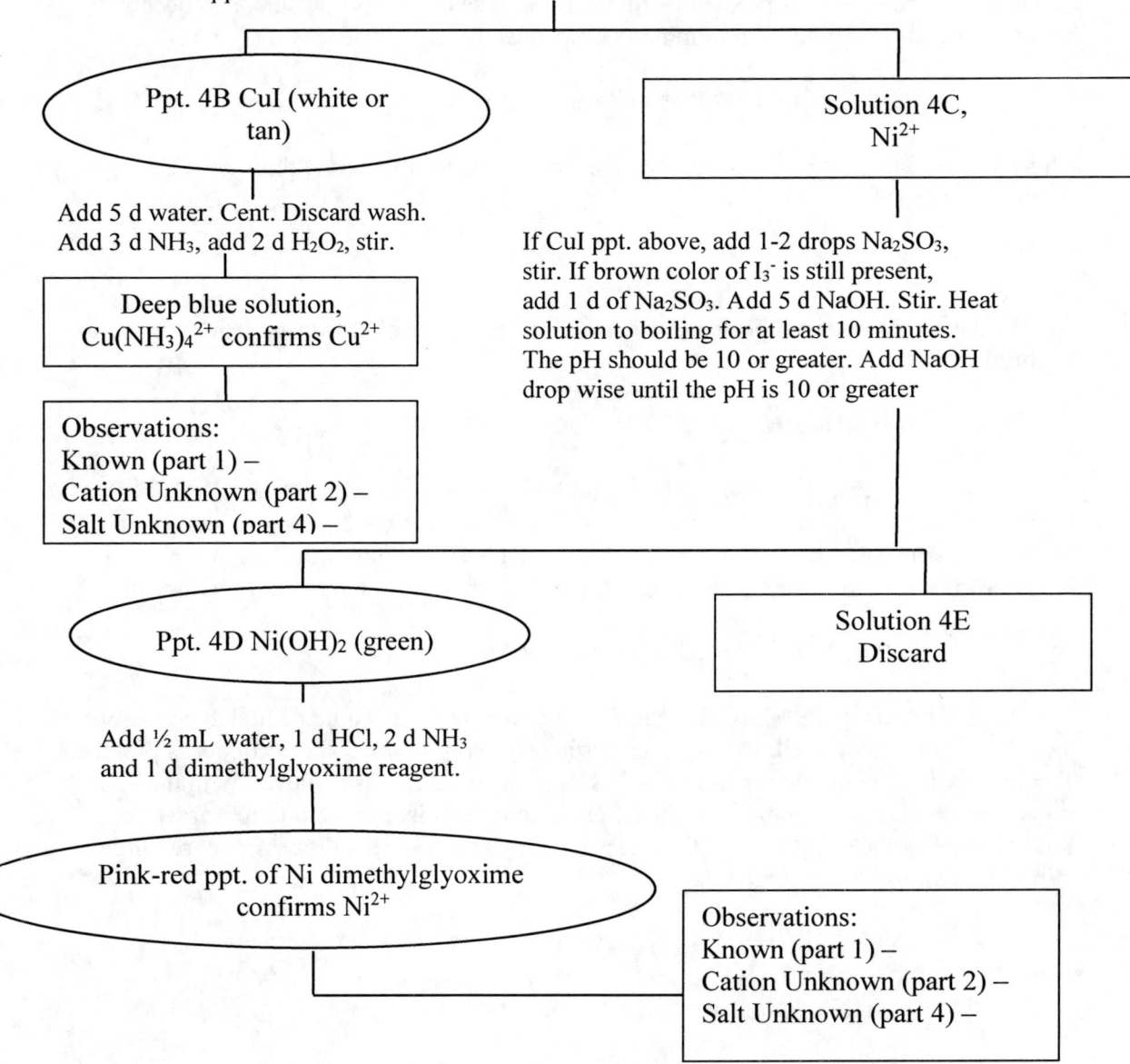

Solution 4A from Group 3
$Cu(NH_3)_4^{2+}$(deep blue), $Ni(NH_3)_6^{2+}$(light violet)

Add 6.0 M HCl until the solution is neutral plus 1 d more. The solution should just turn colorless. Add 3 d KI, stir. Brown color due to I_3^- develops in addition to tan ppt. If Cu^{2+} is present, wait 5 minutes for ppt. to develop. Centrifuge. Test for completeness of ppt. of CuI by adding 1 d KI. Continue until no further ppt. forms.

Ppt. 4B CuI (white or tan)

Solution 4C, Ni^{2+}

Add 5 d water. Cent. Discard wash. Add 3 d NH_3, add 2 d H_2O_2, stir.

Deep blue solution, $Cu(NH_3)_4^{2+}$ confirms Cu^{2+}

Observations:
Known (part 1) –
Cation Unknown (part 2) –
Salt Unknown (part 4) –

If CuI ppt. above, add 1-2 drops Na_2SO_3, stir. If brown color of I_3^- is still present, add 1 d of Na_2SO_3. Add 5 d NaOH. Stir. Heat solution to boiling for at least 10 minutes. The pH should be 10 or greater. Add NaOH drop wise until the pH is 10 or greater

Ppt. 4D $Ni(OH)_2$ (green)

Solution 4E Discard

Add ½ mL water, 1 d HCl, 2 d NH_3 and 1 d dimethylglyoxime reagent.

Pink-red ppt. of Ni dimethylglyoxime confirms Ni^{2+}

Observations:
Known (part 1) –
Cation Unknown (part 2) –
Salt Unknown (part 4) –

Chemical Equilibria and Reactions Involved in the Cation Scheme.

Group 4. Separation and Identification of the Copper Group

The solution remaining from the separation of Mg^{2+} contains $Ni(NH_3)_6^{2+}$ and $Cu(NH_3)_4^{2+}$. These complexes are decomposed by the addition of hydrochloric acid. The separation of copper that follows affords the opportunity of carrying out two oxidation-reduction reactions. The first is the reduction of Cu(II) to Cu(I) using iodide, forming the insoluble tan copper (I) iodide. The brown solution that develops as the copper(I) iodide precipitates results from the formation of the very stable I_3^- complex in solution.

$$I_2 + I^- \rightleftarrows I_3^-$$

The precipitated copper(I) iodide is re-oxidized in the presence of ammonia with hydrogen peroxide, to yield the intense blue $Cu(NH_3)_4^{2+}$ in solution:

$$CuI(s) \rightleftarrows Cu^+ + I^-$$
$$2Cu^+ + H_2O_2 \rightleftarrows 2Cu^{2+} + 2OH^-$$
$$Cu^{2+} + 4NH_3 \rightleftarrows Cu(NH_3)_4^{2+}$$

The sequence of reactions given above are a simplified view of the complex series of reactions that must occur in this case, but the net reaction is correct, production of $Cu(NH_3)_4^{2+}$ from the copper(I) iodide. The intense blue color here is a confirmatory test for copper.

In the second oxidation-reduction titration, the solution containing nickel is titrated with sodium sulfite to remove the elemental iodine produced in the precipitation of copper (I) iodide,

$$SO_3^{2-} + I_3^- + H_2O \rightleftarrows SO_4^{2-} + 3I^- + 2H^+$$

The end point for this reaction, at which drop wise addition of the Na_2SO_3 solution is discontinued, is indicated by the disappearance of the brown color due to the I_3^- ion. Heating the solution after adding NaOH drives off ammonia, allowing nickel to precipitate as the hydroxide.

$$Ni^{2+} + 2OH^- \rightleftarrows Ni(OH)_2(s)$$

The light green ppt. of $Ni(OH)_2$ is dissolved in HCl. The confirmatory test for nickel involves formation of a pink-red chelate complex precipitate.

Part 3. Anion Analysis

The cation analytical scheme outlined above is based on a systematic series of separations, followed by tests to identify individual ions once they have been isolated from the other cations in the scheme. The anion analysis is very different. In this analysis a series of tests will be carried out on each known or unknown sample. More than one anion may give a similar result in any given test, but by considering the results of all the tests, each of the anions included in the scheme for analysis can be distinguished. It may be desirable to carry out confirmatory tests for those anions that are indicated to be present by the results of the spot tests with the general reagents.

The anions included for study in this scheme are: Cl^-, Br^-, I^-, SO_3^{2-}, SO_4^{2-}, PO_4^{3-} and NO_3^-. These anions will be provided in the form of a soluble ionic solid. All of the tests, except the first one with H_2SO_4, will be carried out on water solutions of the solids. Only a small amount of the solid is necessary to make a water solution concentrated enough to produce positive results.

Solutions need for this analysis:

H_2SO_4	18 M	$Ba(C_2H_3O_2)_2$	0.2 M
$AgNO_3$	0.2 M	HCl	2.0 M
HNO_3	3.0 M	$FeSO_4$	0.2 M

Procedure

The series of tests given below are to be performed using a known sample of each anion considered here. The student's observations are to be recorded in Table 1. Record the observations for each test in the appropriate block. Note any color changes in the solid, any gas evolution and the color and odor of the gas evolved, whether or not the solid dissolves, the color of any solution formed, etc. Direct comparisons of tests of any unknown with tests of a known sample are helpful and are encouraged.

Table 1

Ion Tested	Test 1 Conc. H_2SO_4	Test 2 $AgNO_3$	HNO_3	Test 3 Barium Acetate	HCl	Test 4 Special
Cl^-						X
Br^-						X
I^-						X
SO_3^{2-}						X
SO_4^{2-}						X
PO_4^{3-}						X
NO_3^-						
Anion 1 Unknown						
Anion 2 Unknown						
Salt Unknown						

Test 1. Treatment with conc. H_2SO_4. (Safety: Concentrated H_2SO_4 is much more reactive than the dilute solutions used in the cation analysis and will cause immediate painful harm to skin tissue. Your Teaching Assistant will

dispense the concentrated **H₂SO₄ but be careful not to touch it in your test.**) Place a small amount (about 1/8 inch) of the **solid** in a micro test tube. Add 1 to 2 d conc. H₂SO₄ **(your TA will add the H₂SO₄ for you).** Observe and record the results.

CAUTION: Never place your nose **directly over** the mouth of a test tube to determine odors. Gently wave your hand over the top of the test tube, fanning any gas toward your nose. Test each of your solids in this manner and record the results in Table 1.

Preparation of anion solutions. Place the amount of solid that covers approximately 1/8 of the tip of a spatula into a micro test tube and add 2-3 mL of water. If this amount of salt does not all dissolve, centrifuge and decant - using the solution and discarding the undissolved solid. Use portions of this solution in procedures 2 and 3.

Test 2. Treatment of the solution with silver nitrate. Place 2 d of the water solution in a micro test tube. Add 3-4 d of the 0.2 M AgNO₃ solution. Observe and record the results. If a ppt. forms, centr. and decant. Add 10 d of 3.0 M HNO₃ solution to the ppt. Stir and record any results.

Test 3. Treatment of the solution with barium acetate solution. Place 5 d of the solution in a micro test tube. Add 2-3 d of 0.2 M barium acetate solution. Observe and record the results. If a ppt. forms, centr. and decant. Add several drops of HCl to the solid, stir, and note any results.

Prepare solutions and test each of your solids as in 2 and 3 and record the results in Table 1.

Test 4. Special test for nitrate anion. Since all the tests for NO_3^- are negative, the following positive test is recommended. Place 2 d of the solution in a micro test tube. Add carefully 10 d conc. H₂SO₄ **(your TA will add the H₂SO₄ for you).** Mix thoroughly and cool. Carefully add 3-4 d FeSO₄ solution, allowing the latter to float on top of the H₂SO₄ solution. Allow the tube to stand for two minutes. A brown coloration at the junction of the two layers confirms the presence of nitrate anion. The reactions that occur in this test are first reduction of NO_3^- to NO by Fe^{2+} at high [H⁺],

$$NO_3^- + 4H^+ + 3Fe^{2+} \rightarrow NO + 3Fe^{3+} + 2H_2O$$

followed by formation of the brown $FeNO^{2+}$ complex by reaction of NO with unoxidized Fe^{2+}

$$Fe^{2+} + NO \rightarrow FeNO^{2+}$$

Analysis of Unknowns --- Carry out the above tests on each of the unknowns. A comparison of these results with those using the known salts should indicate which anion is present in the unknown. For each unknown (labeled 1 and 2), record on Table 1, with a simple yes or no, if the result on the unknown was the same as with the known. Use this information (a series of yeses) to determine which anion is present in each of your unknowns and place the answer on the answer sheet.

Part 4. Identification of an Unknown Ionic Compound (Salt)

The final analysis is of a pure ionic compound to determine the identity of the cation and anion present. The samples contain only one type of cation and one type of anion, each being a member of the corresponding set of ions already considered in Parts 1, 2, and 3. All the compounds that will be analyzed are soluble in water. Preliminary obser-vations of the solid sample should include noting its color and the pH of the solution prepared for the following analyses.

Analysis for the anion. Carry out the anion analysis just as directed in Part 3, placing a yes/no on the appropriate line of Table 1. Record your result on the answer sheet.

Analysis for the cation. Dissolve 150 to 200 mg of solid sample (a volume equal to about 2 drops of water) in 10 mL of water. Use a 1 mL portion of this solution to analyze for the cation present using the procedures given in Parts 1 and 2. Record your result.

At this point, on the answer sheet write the formula of the compound indicated by your anion analysis and cation analysis. Be careful to remember the charges of the ions so as to get the formula correct. Is the final result a reasonable one? Some years ago, a student reported that his water soluble unknown salt had to be AgCl. In Parts 1, 2 and 3, it has been seen that AgCl is insoluble in water!

- 130 -

Experiment 13 Qualitative Inorganic Analysis

Worksheet 13

Name _____

Date _____ Lab Instructor/Section _____

Qualitative Analysis Answer Sheet: Part 1

I have completed Flowsheets 1-4 with the Lab 13 Cation Known. I have recorded observations (color, precipitation, etc.) in my lab manual for use during the rest of Lab 13.

Signature:_____

To be completed by TA:

Notes of observations for cation known solution (40)   _____

Citizenship/Safety points (40)   _____

Pre-lab (20)   ___20___

Total (out of 100)   _____

Worksheet 13

Name _____

Date _____ Lab Instructor/Section _____

Qualitative Analysis Answer Sheet: Part 2

Group Number: _____

Circle the cations you have identified as present in your unknown:

Ag^+ Fe^{3+} Al^{3+} Ba^{2+} Mg^{2+} Cu^{2+} Ni^{2+}

To be completed by TA:

Results (40)   _____

Citizenship/Safety points (40)   _____

Pre-lab (20)   ___20_____

Total (out of 100)   _____

Worksheet 13

Name _____

Date _____ Lab Instructor/Section _____

Qualitative Analysis Answer Sheet: Part 3

Group Number: _____

Circle the anions you have identified as present in your unknowns:

Unk. 1: Cl^- Br^- I^- PO_4^{3-} SO_4^{2-} SO_3^{2-} NO_3^-

Unk. 2: Cl^- Br^- I^- PO_4^{3-} SO_4^{2-} SO_3^{2-} NO_3^-

To be completed by TA:

Unknown 1 Result (20)   _____

Unknown 2 Result (20)   _____

Citizenship/Safety points (40)   _____

Pre-lab (20)   ___20___

Total (out of 100)   _____

Experiment 14 Molar Mass Determination by Freezing Point Depression

Worksheet 13

Name_____

Date_____ Lab Instructor/Section_____

Qualitative Analysis Answer Sheet: Part 4

Group Number:_____

Cation: Ag^+ Fe^{3+} Al^{3+} Ba^{2+} Mg^{2+} Cu^{2+} Ni^{2+}

Anion: Cl^- Br^- I^- PO_4^{3-} SO_4^{2-} SO_3^{2-} NO_3^-

The chemical formula for my ionic compound is_____.

To be completed by TA:

Cation Result (20)   _____

Anion Result (20)   _____

Citizenship/Safety points (40)   _____

Pre-lab (20)   ___20___

Total (out of 100)   _____

Experiment 14

Molar Mass Determination by Freezing Point Depression

Objective

Calculate the molar mass of an unknown solid from the freezing point depression.

Equipment

CBL system
TI graphing calculator
Vernier temperature probe
Thermometer
Ring stand
150-mL beaker

Utility clamp
18 X 150-mm test tube
Dodecanol
Unknown solid
2- 400-mL beakers

Safety Precautions

Handle the dodecanol with care. Be careful with open flames.

Principles

When a non-volatile solute is dissolved in a pure liquid the physical properties (i.e. boiling point, freezing point, melting point etc.) of the liquid change. The degree to which the properties change is directly related to the concentration of the solute particles present. Properties of a solution that change in proportion to the concentration of solute particles are called **colligative** properties. Colligative properties do not depend on the nature of solute particles, only on their concentrations. Colligative properties are a means of determining the number of solute particles present in a particular sample – data that can be used to determine the molar mass of the solute.

Experiment 14 — Molar Mass Determination by Freezing Point Depression

For example, when a solute is dissolved in a pure liquid the freezing temperature of the solution is lowered in proportion to the number of moles of solute added. This *colligative property* is known as freezing-point depression. The equation that shows this relationship is:

$$\Delta T = i\, m\, k_f \qquad (1)$$

where ΔT is the freezing point depression, k_f is the freezing point depression constant for a particular solvent (2.9°C-kg/mol for dodecanol in this experiment), i is the v'ant Hoff factor (number of ions formed when one formula unit of the solid dissolves) and m is the molality of the solution (in mol solute/kg solvent).

In this experiment, you will first find the freezing temperature of the pure solvent, dodecanol. You will then add a known mass of an unknown solute (a non-electrolyte), to a known mass of dodecanol, and determine the lowering of the freezing temperature of the solution. By measuring the freezing point depression, ΔT, and the mass of the unknown, you can calculate the molar mass of the unknown solute, in g/mole. For example, if the freezing point of pure dodecanol was determined to be 27.5 °C (***Note. This is not actual freezing point; you will determine it yourself in the course of the experiment***) and after dissolving 1.32 grams of an unknown solid (non-electrolyte) to 12.5 grams of the dodecanol the freezing point was lowered to 25.4 °C. The molar mass of the unknown can be determined using

$$\text{Molar mass} = \frac{g}{mol} = \frac{1.32\ g\ solute}{moles\ solute}$$

with moles of solute being determined from molality found in Equation 1.

$$m = \frac{\Delta T}{i \cdot k_f} \quad \text{where,}$$

$$\Delta T = (27.5\ °C - 25.4\ °C) = 2.1\ °C$$

$$k_f = 2.9\ °C\text{-kg/mol} \qquad i = 1$$

$$m = \frac{2.1\ °C}{(1) \cdot (2.9\ °C\text{-kg}/mol)} = 0.72\ \text{molal}$$

From this calculation, we know there is 0.72 mole of solute in every 1 kg of solvent.

Moles solute =

$$12.5g \left(\frac{1\ kg}{1000\ g} \right) \left(\frac{0.72\ mol}{1\ kg\ solvent} \right)$$

$$= 0.0090\ mol$$

Now, insert this into the molar mass equation shown previously.

$$\text{Molar mass} = \frac{g}{mol} = \frac{1.32\ g\ solute}{0.0090\ moles\ solute}$$

$$= 1.5 \times 10^2\ g/mol$$

PROCEDURE

1. Work in pairs
   - Wear safety glasses
   - Obtain from your TA a CBL system

2. Plug the temperature probe into the adapter cable in Channel 1 of the CBL System. Use the link cable to connect the CBL System to the TI Graphing Calculator. Firmly press in the cable ends. Make sure the CBL power cable is plugged in.

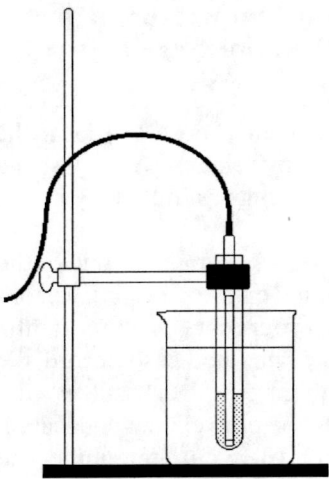

Figure 1

3. Turn on the CBL unit and the calculator. Start the CHEMBIO program and proceed to the MAIN MENU.

4. Set up the calculator and CBL for one temperature probe and a temperature calibration.

   - Select SET UP PROBES from the MAIN MENU.
   - Enter "1" as the number of probes.
   - Select TEMPERATURE from the SELECT PROBE menu.
   - Enter "1" as the channel number.

5. Set up the calculator and CBL for data collection.

   - Select COLLECT DATA from the MAIN MENU.
   - Select TIME GRAPH from the DATA COLLECTION menu.
   - Enter "6" as the time between samples, in seconds.
   - Enter "99" as the number of samples (the CBL will collect data for a total of 9.9 minutes).
   - Press [ENTER]. Select USE TIME SETUP to continue. If you want to change the sample time or sample number, select MODIFY SETUP.
   - Enter "0" as the minimum temperature (Ymin).
   - Enter "100" as the maximum temperature (Ymax).
   - Enter "5" as the temperature increment (Yscl).
   - Do not begin data collection until directed to do so in Step 8.

Part I Freezing Temperature of Pure Dodecanol

6. Add about 300 mL of tap water to a 400-mL beaker. Place the beaker on the base of the ring stand and using a Bunsen burner bring the water to a boil. **(After the water starts to boil turn the flame off)**

7. Place the test tube into an empty 150-mL beaker (for support) record the mass of the empty test tube and beaker in your data section.

8. Fill the test tube about ¼ full with dodecanol place the test tube back in the 150-mL beaker from step 7 and reweigh and record the mass in your data section.

9. Prepare an ice bath by filling the second 400-mL beaker with ice and a small amount of water.

10. Insert the alcohol thermometer into the test tube containing the solvent and place the test tube in the hot water bath prepared in step 1. **(Be sure the flame is off).** Allow the temperature of the solvent to rise to about 45-55 °C, and then remove it from the hot water bath. Remove the thermometer.

11. Insert the temperature probe into the hot dodecanol. About 30 seconds are required for the probe to warm up to

the temperature of its surroundings and give correct temperature readings. During this time, fasten the utility clamp to the ring stand so the test tube is above the ice water bath. After the 30 seconds have elapsed, press ENTER on the calculator to begin data collection.

12. Lower the test tube into the ice water bath. Make sure the water level outside the test tube is higher than the dodecanol level inside the test tube, Figure 1.

13. With a very slight up and down motion of the temperature probe, *continuously* stir the dodecanol during the cooling. Hold the top of the probe and *not* its wire.

14. Continue with the experiment until data collection has stopped after 9.9 minutes (when "DONE" appears on the CBL screen). Use the hot water bath to melt the dodecanol before pulling the probe out of the now liquid dodecanol. *Do not* attempt to pull the probe out of the solid for this might damage it. Carefully wipe any excess dodecanol liquid from the probe with a paper towel or tissue. Return the test tube containing dodecanol to the 150-mL beaker for storage until needed in the following steps.

15. Press ENTER to display a graph of temperature vs. time on the calculator screen. To determine the freezing temperature of pure dodecanol, you need to determine the temperature in the portion of the graph with nearly constant temperature. Examine the data points along this portion of the graph. As you move the cursor right or left, the time (X) and temperature (Y) values of each data point are displayed below the graph. Determine this temperature, either by visually approximating the value or by taking the mathematical average of the temperatures in this plateau. Record the freezing temperature of pure dodecanol in your data table (round to the nearest 0.1°C).

Part 2 Freezing Temperature of a Solution of an unknown and Dodecanol

16. Press ENTER, then choose to repeat the data collection by selecting YES. Use the same Y-axis settings as in Part I.

17. Weigh about 1.0-1.5 grams of one of the unknowns found on the sideshelf. Record the mass in your data section to the nearest 0.01 grams and also record the unknown number. Dissolve the unknown into the preweighed dodecanol. Repeat Steps 10-15 to determine the freezing point of this mixture.

Processing the Data

1. Determine the difference in freezing temperatures, ΔT, between the pure dodecanol (T_1) and the mixture of dodecanol and the unknown solid (T_2). Use the formula, $\Delta T = T_1 - T_2$.

2. Calculate molality (m), in mol/kg, using the formula, $\Delta T = imk_f$ (k_f = 2.9°C-kg/mol for dodecanol).

3. Calculate moles of the unknown solute, using the method described in the discussion.

4. Calculate the molar mass of the unknown, in g/mol.

5. From the chart on the next page determine the identity of the unknown.

Potential solutes	Molar mass (g/mol)
Benzoic acid	122
Fructose	180
Citric acid	192
Palmetic acid	256
Steric acid	285

Experiment 14 — Molar Mass Determination by Freezing Point Depression

Worksheet 14

Name _____

Date _____ Lab Instructor/Section _____

Pre-lab	____/20
Data	____/20
Post-lab	____/20
Safety/Part.	____/40
Total	____/100

Data sheet

1. Mass of empty test tube and beaker   _____ g

2. Mass of filled test tube and beaker   _____ g

3. Mass of dodecanol (2-1)   _____ g

4. Mass of dodecanol in kg   _____ kg

5. T_F for pure dodecanol   _____ °C

6. Mass of unknown   _____ g

7. T_F for solution   _____ °C

Calculations (show work)

Identity of unknown   _____

Experiment 14   Molar Mass Determination by Freezing Point Depression

Experiment 2 Questions

1. Your calculation on the data sheet was for a non-electrolyte ($i = 1$). Enter the molar mass you calculated on the data table in the labeled box below.

   You learn the sample ionizes into two particles. In the space below, recalculate the molar mass of the sample using this new information and enter that answer in the labeled box.

Molar mass from Data Sheet:

Molar mass from re-calcualtion:

2. Would the calculated molar mass be higher, lower or the same as the actual molar mass of the compound if the unknown contained an insoluble impurity? Explain your answer for full credit.

Experiment 14 Molar Mass Determination by Freezing Point Depression

Pre-Laboratory 14

Name _____

Date _____ Lab Instructor/Section _____

Pre-laboratory

1. Calculate the molar mass of an unknown solid (assume a non-electrolyte), if 1.35 grams of the solid was dissolved in 15.0 grams of dodecanol and caused a freezing point depression of 2.34 °C. The freezing point constant, k_f, is in the lab discussion.

2. Using the information in #1, calculate the molar mass of this compound if it were actually an ionic compound that dissociates into two parts. (i.e. BX → B$^+$ + X$^-$)

Experiment 15

Diprotic Acids: Identifying an Unknown by Titration

Objective

Identify an unknown diprotic acid by using a titration to determine its molar mass.

Equipment

CBL system
TI graphing calculator
AC adapter
Vernier pH amplifier and pH electrode
Vernier adapter cable
0.1 M NaOH solution (standardized)

250-mL beaker
Stirring rod
Utility clamps
Distilled water

Safety Precautions

Sodium hydroxide solution is caustic. Avoid spilling it on your skin or clothing.

Handle the acid, in both the solid state and its solution, with care. Acids can harm your eyes, skin, and respiratory tract.

Principles

A diprotic acid is one that contains two acidic hydrogen ions (H$^+$) per molecule of acid. Examples of diprotic acids are sulfuric acid, H_2SO_4, and carbonic acid, H_2CO_3. Diprotic acids ionize in water in two steps:

$$H_2X_{(aq)} \rightleftarrows H^+_{(aq)} + HX^-_{(aq)} \quad (1)$$

$$HX^-_{(aq)} \rightleftarrows H^+_{(aq)} + X^{2-}_{(aq)} \quad (2)$$

Because it is more difficult to remove the proton from the anionic HX$^-$ species than from neutral H_2X, the titration curve for diprotic acids reacting with a base has two equivalence points, as shown in Figure 1.

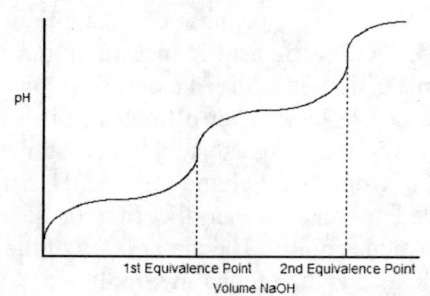

Figure 1

- 149 -

Analogous to the two ionization steps, two equations represent the acid-base reactions occurring between a diprotic acid, H_2X, and sodium hydroxide, NaOH.

From the start of the titration to the first equivalence point:

$$H_2X + NaOH \rightarrow NaHX + H_2O \qquad (3)$$

From the first to the second equivalence point:

$$NaHX + NaOH \rightarrow Na_2X + H_2O \qquad (4)$$

From the start of the reaction through the second equivalence point the net reaction is:

$$H_2X + 2NaOH \rightarrow Na_2X + 2H_2O \qquad (5)$$

From equation 3, we see that at the first equivalence point all of the hydrogen ions from the first ionization reaction have reacted with NaOH. From equation 4, at the second equivalence point all of the H^+ ions from both reactions have reacted with the NaOH. It should be noted that at the second equivalence point exactly twice as many H^+ ions have reacted with the NaOH as did at the first equivalence point. Therefore, the volume of NaOH added at the second equivalence point is exactly twice that of the first equivalence point.

The primary purpose of this experiment is to identify an unknown diprotic acid by finding its molar mass. The mass of a sample of an unknown diprotic acid is measured and this sample is titrated with a standardized solution of NaOH. The number of moles of the acid present in the weighed sample can be determined from the volume of NaOH titrant needed to reach either the first or second equivalence point. The molar mass of the diprotic acid is then found in g/mol:

$$\text{Molar mass of the acid} = \frac{\text{grams of unknown acid}}{\text{moles of unknown acid}} \qquad (6)$$

In the experiment, we will use the titration curve that is more clearly defined to calculate the moles of acid. If the first equivalence point is used, equation 3, the moles of NaOH equal the moles of acid. If the second equivalence point is used, equation 5, the conversion factor

$$\frac{1 \text{ mole } H_2X}{2 \text{ mole NaOH}}$$

must be used. For example, if it were known that exactly 25 mL of a 0.10 molar solution of NaOH was needed to reach the second equivalence point during the titration of an unknown diprotic acid:

$$0.025 \text{ L NaOH} \times \left(\frac{0.10 \text{ moles NaOH}}{1 \text{ Liter of soluiton}}\right)\left(\frac{1 \text{ mole } H_2X}{2 \text{ moles NaOH}}\right)$$

$$= 0.0010 \text{ moles of } H_2X$$

In this experiment you will be given one of five possible diprotic acids. You will use the technique outlined above to determine the molar mass of the acid and use this information to identify the acid.

Experiment 15 Diprotic Acids: Identifying an Unknown by Titration

Procedure

1. Work in pairs
   - Wear safety glasses.
   - Obtain a CBL system from your TA

2. Weigh out about 0.12 g of the unknown diprotic acid on a piece of weighing paper. Record the mass to the nearest 0.01 g in the Data and Calculations table. Transfer the unknown acid to a 250-mL beaker and dissolve it in about 100 mL of distilled water. **CAUTION:** *Handle the solid acid and its solution with care. Acids can harm your eyes, skin, and respiratory tract.*

3. Place the beaker on a magnetic stirrer and add a stirring bar. If no magnetic stirrer is available, you need to stir with a stirring rod during the titration.

4. Prepare the pH system for data collection.

   - Plug the pH amplifier into the adapter cable in Channel 1 of the CBL System. The pH electrode is already connected to the pH amplifier.
   - Use the link cable to connect the CBL System to the TI Graphing Calculator. Firmly press in the cable ends.

5. Use a utility clamp to suspend a pH electrode on a ring stand as shown in Figure 2. Position the pH electrode in the acid solution and adjust its position toward the outside of the beaker.

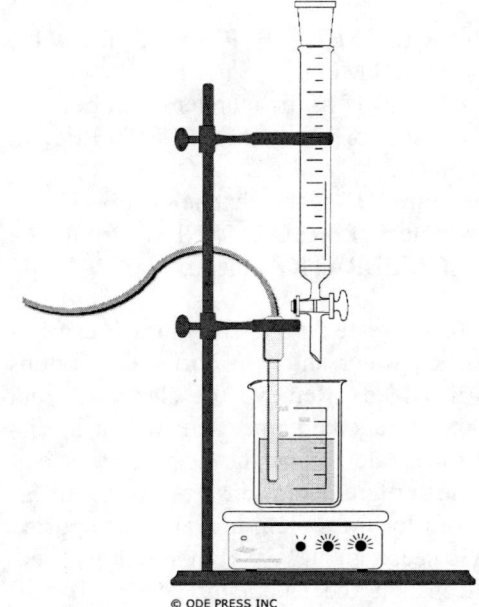

Figure 2

6. Obtain a 50-mL buret and rinse the buret with a few mL of the ~0.1 M NaOH solution. Record the precise concentration of the NaOH solution in the Data and Calculations table. Use a utility clamp to attach the buret to the ring stand as shown in Figure 2. Fill the buret a little above the 0.00-mL level of the buret. Drain a small amount of NaOH solution so it fills the buret tip *and* leaves the NaOH at the 0.00-mL level of the buret. Dispose of the waste solution in this step as directed by your teacher. **CAUTION:** *Sodium hydroxide solution is caustic. Avoid spilling it on your skin or clothing.*

7. Turn on the CBL unit and the calculator. Start the CHEMBIO program and proceed to the MAIN MENU.

8. Set up the calculator and CBL for pH measurement.

- Select SET UP PROBES from the MAIN MENU.
- Enter "1" as the number of probes.
- Select pH from the SELECT PROBE menu.
- Enter "1" as the channel number.
- Select PERFORM NEW from the CALIBRATION menu.

To calibrate the system you will use distilled water and two buffered solutions (provided). Remove the electrode from the storage container and rinse it in distilled water. Place the electrode in one of the buffered solutions (e.g., 4.0 buffer) wait for the system to warm up (approx. 30 seconds) hit [ENTER] then follow directions on the calculator screen. It will prompt you for calibration point 1. When the CBL has stabilized push [TRIGGER] on the CBL. The calculator will then prompt you to enter a reference, when using a 4.0 buffer you should put 4.0 as your reference. After entering 4.0 the calculator will prompt you to calibrate point 2. The electrode should be removed from the 4.0 buffer rinsed in distilled water and placed into the 7.0 buffer and perform the same tasks as above. When finished the calculator will prompt to push [ENTER] and move on to the next step.

9. Set up the calculator and CBL for data collection.

    - Select COLLECT DATA from the MAIN MENU.
    - Select TRIGGER/PROMPT from the DATA COLLECTION menu. Follow the directions on the calculator screen to allow the system to warm up, then press [ENTER].

10. You are now ready to begin the titration. This process goes faster if one person manipulates and reads the buret while another person operates the calculator and enters buret readings.

    - Press [ENTER] after the pH system has warmed up 30 seconds. Before adding NaOH titrant, monitor the pH value on the CBL screen. Once the pH has stabilized, press [TRIGGER] on the CBL and enter "0" (the buret volume, in mL) in the TI-83 calculator. You have now saved the first data pair for this experiment.
    - Select MORE DATA to collect another data pair. Add enough NaOH to raise the pH by about 0.20 units. After the NaOH has been added and the pH stabilizes, press [TRIGGER] and enter the current buret reading. You have now saved the second data pair for the experiment, to the nearest 0.01-mL.
    - Select MORE DATA and continue adding NaOH solution in increments that raise the pH about 0.20 units. Enter the buret reading after each addition. Proceed in this manner until the pH is 3.5.
    - When pH 3.5 is reached, change to 2-drop increments. Enter the buret reading after each increment.
    - After pH 4.5 is reached, again add larger increments that raise the pH by about 0.20 units and enter the buret reading after each addition. Continue in this manner until a pH of 7.5 is reached.
    - When pH 7.5 is reached, change to 2-drop increments. Enter the buret reading after each increment.
    - When pH 10 is reached, again add larger increments that raise the pH by 0.20 units. Enter the buret reading after each increment. Continue in this manner until you reach a pH of 11 or use 25 mL of NaOH, whichever comes first.

11. Select STOP AND GRAPH from the DATA COLLECTION menu when you have finished collecting data. Examine the data points along the displayed graph of pH vs. NaOH volume. As you move the cursor right or left, the volume (X) and pH (Y) are displayed below the graph.

One of the two equivalence points is usually more clearly defined than the other; the two-drop increments near the equivalence points frequently result in larger increases in pH (a steeper slope) at one equivalence point than the other. Indicate the more clearly defined equivalence point (first or second) in Box 1 of the Data and Calculations table. Determine the volume of NaOH titrant used for the equivalence point you selected. To do so, examine the data to find the largest increase in pH values during the 2-drop additions of NaOH. Find the NaOH volume just *before* this jump. Then find the NaOH volume *after* the largest pH jump. Record these values in Box 2 of your data table.

For the *alternate* equivalence point (the one you did *not* use in the previous step), examine the data points on your graph to find the largest increase in pH values during the 2-drop additions of NaOH. Find the NaOH volume just *before* and *after* this jump. Record these values in Box 10 of your data table.

12. Dispose of the beaker and buret contents as directed by your teacher. Rinse the pH electrode with distilled water and return it to the storage solution.

Processing the Data

1. Use your graph and data table to confirm the volumes you recorded in Box 2 of the Data and Calculations table (volumes of NaOH titrant *before* and *after* the largest increase in pH values).

2. Determine the volume of NaOH added at the equivalence point you selected in Step 1. To do this, add the two NaOH volumes determined in Step 1, and divide by two.

3. Calculate the number of moles of NaOH used at the equivalence point you selected in Step 1.

4. Determine the number of moles of the diprotic acid, H_2X. Use Equation 3 or Equation 5 to obtain the ratio of moles of H_2X to moles of NaOH, depending on which equivalence point you selected in Step 1.

5. Using the mass of diprotic acid you measured in Step 1 of the procedure, calculate the molar mass of the diprotic acid, in g/mol.

6. From the following list of five diprotic acids, identify your unknown diprotic acid.

Diprotic Acid	Formula	Molar mass (g/mol)
Oxalic Acid	$H_2C_2O_4$	90
Malonic Acid	$H_2C_3H_2O_4$	104
Maleic Acid	$H_2C_4H_2O_4$	116
Malic Acid	$H_2C_4H_4O_5$	134
Tartaric Acid	$H_2C_4H_4O_6$	150

7. Determine the percent error for your molar mass value in Step 5.

8. Use your graph and data table to confirm the volumes you recorded in Box 10 of the Data and Calculations table (volumes of NaOH titrant *before* and *after* the largest increase in pH values at the alternate equivalence point). Note: Dividing or multiplying the other equivalence point volume by two may help you confirm that you have selected the correct two data pairs in this step.

9. Determine the volume of NaOH added at the alternative equivalence point, using the same method you used in Step 2 of Processing the Data.

Experiment 15   Diprotic Acids: Identifying an Unknown by Titration

Worksheet 15

Name _____

Date _____ Lab Instructor/Section _____

Pre-lab	____ /20
Data	____ /20
Post-lab	____ /20
Safety/Part.	____ /40
Total	____ /100

Data sheet

Mass of Diprotic acid _____ g   Concentration of NaOH _____ M

1. Equivalence point (indicate the one you will use in the calculations below)	first equivalence point ____ or second equivalence point ____
2. NaOH volume added before and after largest pH increase	_____ mL _____ mL
3. Volume of NaOH added at the equivalence point	_____ mL
4. Moles NaOH	_____ mol
5. Moles of diprotic acid, H_2X	_____ mol
6. Molar mass of the diprotic acid	_____ g/mol
7. Name, formula, and accepted molar mass of the diprotic acid	_____ _____ _____ g/mol
8. Percent error	_____ %
9. Alternate equivalence point (indicate the one used below)	first equivalence point ____ or second equivalence point ____
10. NaOH volume before and after largest pH increase (alternate equivalence point)	_____ mL _____ mL
11. Volume of NaOH added at the alternate equivalence point	_____ mL

Questions

1. What effect if any would the following have on the molar mass of the acid

   a. A non acidic, soluble impurity contaminated the acid weighed out in this experiment.

   b. The concentration of the NaOH was actually twice as concentrated as reported.

   c. The titrant was not NaOH but $Ca(OH)_2$

Experiment 15   Diprotic Acids: Identifying an Unknown by Titration

Pre-Laboratory 15

Name _____

Date _____   Lab Instructor/Section _____

Pre-laboratory

1. Write the chemical equation for the reaction of H_2SO_4 with LiOH, forming Li_2SO_4. Calculate the number of moles of H_2SO_4 needed to react with 25.0 mL of 0.20 M LiOH.

2. What is the molar mass of an unknown diprotic acid if 65.0 mL of a 0.15 M solution of KOH was needed to reach the second equivalence point in the titration of 0.30 grams of the acid?

Experiment 16

The Acid Ionization Constant (K_a) of Acetic Acid

Objective

Calculate the acid ionization constant of acetic acid by measuring the pH of various concentrations of acetic acid / sodium acetate solutions.

Equipment and Chemicals

2-Burets
pH paper 3-ranges
0.5 M acetic acid
0.5 M sodium acetate
5-test tubes 16 x 150

Safety Precautions

Wear approved eye protection. Avoid spilling chemicals.

Principles

The Arrhenius definition of an acid is a substance that produces hydrogen ions in aqueous solutions.

$$\text{HA (aq)} \rightleftarrows \text{H}^+\text{(aq)} + \text{A}^-\text{(aq)} \tag{1}$$

For several acids equation 1 proceeds essentially to completion, i.e., the entire compound is present in solution as H^+ and A^-. In these cases, the concentration of hydrogen ion is equal to the molar concentration of the compound and the acid is called a strong acid. For a number of other acidic compounds, it is found that only a fraction of the molecules in solution have ionized. These are called weak acids. Experimentally, it has been shown that the extent of ionization of weak acids must satisfy the equation:

$$K_a = \frac{[\text{H}^+][\text{A}^-]}{[\text{HA}]} \tag{2}$$

where K_a is the acid ionization constant, which is a constant for any given acid. HA and A⁻ are related to one another as **conjugate** acid base pairs. The formulas in brackets are the molar concentrations of each of these species in solution. Equation 2 may be rewritten by taking the logarithms of both sides.

$$\log K_a = \log [H^+] + \log \frac{[A^-]}{[HA]} \quad (3)$$

$$\text{or} \quad -\log [H^+] = -\log K_a + \log \frac{[A^-]}{[HA]} \quad (4)$$

Since the left hand side of equation 4 is the definition of pH and $-\log K_a$ is the definition of pK_a we could rewrite equation 4 to read:

$$pH = pK_a + \log \frac{[A^-]}{[HA]} \quad (5)$$

Equation 5 is formally known as the **Henderson Hasselbalch** equation. Since the last term is a ratio of the concentrations of the conjugate base to weak acid we can express this in terms of moles (or millimoles) of base over moles (or millimoles) of acid.

$$pH = pK_a + \log \frac{n_B}{n_A} \quad (6)$$

Equation 6 can then be rearrange solving for pK_a:

$$pK_a = pH - \log \frac{n_B}{n_A} \quad (7)$$

Therefore, by measuring the pH of a solution made by mixing known concentrations of a weak acid, such as formic acid (HCOOH), with its conjugate base, where the base is normally found as a sodium salt (i.e. sodium formate – NaHCOO), the pK_a can be determined.

For example, if the pH of a solution made by mixing 50.0 mL of a 0.350 M formic acid with 35.0 mL of 0.450 M sodium formate was determined to be 3.70 the pK_a would be:

Use Equation 7 where:

n_A (millimoles) =

$$50.0 \text{ mL} \times \frac{0.350 \text{ millimoles}}{1 \text{ mL}}$$

= 17.5 millimol of acid

n_B (millimoles) =

$$35.0 \text{ mL} \times \frac{0.450 \text{ millimoles}}{1 \text{ mL}}$$

= 15.8 millimol of conjugate base

$$pK_a = 3.70 - \log \frac{15.8}{17.5} = 3.74$$

In this experiment a series of solutions having various ratios of sodium acetate to acetic acid are prepared by mixing solutions of acetic acid and sodium acetate of known concentrations. You will then measure the pH of each solution and calculate the pK_a. These values will then be used to calculate the average pK_a for acetic acid and compare it to its actual value.

Procedure

Two solutions will be used in this experiment. One is a 0.50 M solution of sodium acetate (NaOAc) and the other is 0.50 M acetic acid (HOAc). Clean two burets and label one "ACETIC ACID" and the other "SODIUM ACETATE" or appropriate abbreviations. Rinse each buret with 5 mL of the appropriate solution and fill them.

Label a set of test tubes from 1 through 5. From the burets, place into each of the five test tubes the approximate quantities listed in the table in the Data Section, and record your values to two decimal places. Stir the solutions in the test tubes with a clean stirring rod - **be sure to clean and thoroughly dry the rod before using it in a second solution.**

Determine the approximate pH of each solution using broad-range (1-14) pH paper. To accomplish this, obtain a small piece of pH paper about 1/4 " in length and place it on the edge of a clean watch glass. Dip into your test tube with a clean stirring rod and place a drop on the pH paper. Compare the color of the paper with the color chart on the side of pH paper container. This gives an approximate pH. To get a more accurate pH, choose short-range pH paper near the approximate pH and re-test the pH. You should be able to get a pH with a decimal place. If the color appears to match one of the extremes of the range, i.e., either the highest or lowest pH number, then obtain another small piece of pH paper for testing which covers the next higher or lower range as appropriate. Do this until an appropriate pH reading is obtained for each solution and enter the reading in the data table. **Use pH paper sparingly**.

From the known volumes and concentrations calculate the number of millimoles of NaOAc and HOAc used, for each series, and enter the values in the Data table. Use this information to calculate the pK_a for each series.

Experiment 16   The Acid Ionization Constant (*Ka*) of Acetic Acid

Worksheet 16

Name _____

Date _____ Lab Instructor/Section _____

Pre-lab	____/20
Data	____/20
Post-lab	_20_/20
Safety/Part.	____/40
Total	____/100

Data sheet

Test tube	mL of HOAc	mL of NaOAc	pH	Millimoles HOAc	Millimoles NaOAc	pK_a
1	(≈1)	(≈9)				
2	(≈3)	(≈7)				
3	(≈5)	(≈5)				
4	(≈7)	(≈3)				
5	(≈9)	(≈1)				

Average calculated pK_a _____

Calculations:

- 163 -

Experiment 16　　　　　　　　　　　　The Acid Ionization Constant (*Ka*) of Acetic Acid

Pre-Laboratory 16

Name _____

Date _____　Lab Instructor/Section _____

Pre-laboratory

1. During an experiment, a student mixed 14.5 mL of a 0.500 *M* sodium fluoride solution with 15.6 mL of a 0.750 *M* hydrofluoric acid solution the measured pH was 3.24. Calculate the pK_a.

2. Calculate the pH of a solution prepared by mixing 20.5 mL of a 0.500 *M* sodium flouride solution with 32.5 mL of a 0.750 *M* hydrofluoric acid solution. Use the pK_a calculated from question 1.

- 165 -

Experiment 17

Identification of an Unknown Acid By Titration

Objective

An unknown monoprotic acid will be titrated with a standard base solution using CBL technology to record pH and volume measurements. The unknown acid will be identified by comparing the pK_a and molar mass calculated from the titration data with a list of possible acids.

Equipment

CBL system
AC adapter
Vernier adapter cable
TI graphing calculator
Vernier pH amplifier and pH electrode
Butterfly clamp
Buret
Funnel

100-mL beaker
150-mL beaker
Stirring rod
pH 4.0 buffer
pH 7.0 buffer
0.1000 M NaOH solution
0.2 g unknown acid
2 drops 1% phenolphthalein indicator
Distilled water

Safety Precautions

Sodium hydroxide solution is caustic. Avoid spilling it on your skin or clothing. It is far more dangerous to your eyes than is acid.

Handle the acid and the acid solutions you prepare with care. Acids can harm your eyes, skin, and respiratory tract.

Always wear safety glasses/goggles and closed-toe shoes in lab.

Discussion

In solution, weak acids partially ionize forming an equilibrium system. An example of a weak acid is acetic acid. The ionization and equilibrium expression for the ionization of any monoprotic weak acid are provided below.

$$HA(aq) + H_2O(l) \rightleftharpoons H_3O^+(aq) + A^-(aq)$$

$$K_a = \frac{[H_3O^+][A^-]}{[HA]}$$

During the titration of a weak acid, reaction of the weak acid with strong base produces the conjugate base of the weak acid. When the strong base is the limiting reactant, a buffer is produced from the mixture of excess weak acid and produced conjugate base. This area in the titration curve is called the buffer region. Another important landmark on a titration curve is the equivalence point. The equivalence point is defined as the point in the titration when the mole amount of titrant added is equal to the mole amount of the substance being titrated. The equivalence point in the titration of an unknown acid can be determined by finding the inflection point in the graph or looking for the largest change in pH per volume change when the titration has been correctly performed with many meas-

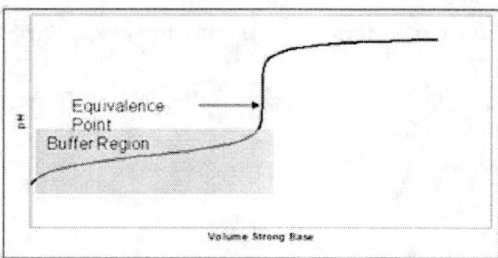

urements in the equivalence point region. An example titration curve for a weak acid with strong base is provided above.

The molar mass of an unknown acid can be determined through calculations using the equivalence point. At the equivalence point, the number of moles of strong base added is equal to the number of moles of unknown acid. Using the volume of base added and the molarity of the base solution, moles of base added can be calculated. Because the unknown is a monoprotic acid, the stoichiometry is known to be 1:1 based on the reaction (1) given below.

$$HA(aq) + NaOH(aq) \rightarrow NaA(aq) + H_2O(l) \quad (1)$$

Once the moles of unknown acid have been calculated, the molar mass can be calculated from the mass of unknown acid and the number of moles, as determined by the titration.

For example, if 15.00 mL of $0.1000 M$ NaOH were required to titrate 0.110 g of unknown acid, the molar mass of the unknown acid would be 73.3 g/mol, as shown in the following calculation:

$$\text{amount of base} = 0.015 \, L \times \frac{0.100 \, moles}{1.00 \, L}$$

$$= 1.500 \times 10^{-3} \, moles \text{ of NaOH}$$

Since the acid is known to be monoprotic, at the equivalence point the number of moles of acid and base are equal.

$$= 1.500 \times 10^{-3} \, moles \text{ of HA}$$

Molar Mass =

$$\frac{0.110 \text{ g acid}}{1.500 \times 10^{-3} \text{ mol acid}} = 73.3 \text{ g/mol}$$

In order to determine the pK_a of an unknown acid, measurements in the buffer region are used. The pH and pK_a of a buffer are related through the Henderson-Hasselbach equation (2). Note: There are some basic assumptions that are required to use the Henderson-Hasselbach equation. This experiment has been designed to meet those assumptions. Refer to your textbook and lecture material for specifics on the use of the Henderson-Hasselbach equation.

$$pH = pKa + \log\frac{N_b}{N_a} \qquad (2)$$

When an equal amount of weak acid and conjugate base exist in solution, the last term in the Henderson-Hasselbach equation becomes zero, and the pK_a equals the pH. This occurs at the half-point to the equivalence point. At the half-point to the equivalence point, enough strong base has been added to react exactly half of the unknown acid and therefore produce an equal amount of conjugate base.

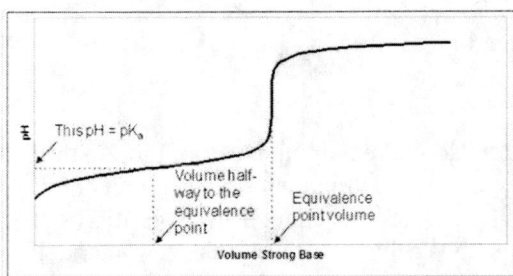

For example, from the following set of data, the pK_a of an unknown acid can be determined. The volume half-way to the equivalence point is 25.00/2 or 12.50 mL. The pH at this point is 4.74. Since the last term in the Henderson-Hasselbach equation is zero, the pK_a is equal to 4.74.

Volume of Base added	Measured pH	
0.00 mL	2.87	
6.25 mL	4.24	
12.50 mL	4.74	
18.75 mL	5.24	
25.00 mL	8.72	←Equivalence Point
31.25 mL	12.03	
37.50 mL	12.31	
43.75 mL	12.44	
50.00 mL	12.52	

In this experiment, you will be assigned an unknown acid to titrate with a standardized sodium hydroxide solution. Phenolphthalein indicator will be added to the unknown acid solution to serve as a visual indicator of the equivalence point. This visual indication of the equivalence point is called the endpoint. Phenolphthalein is colorless in acidic solutions and pink in basic solutions. In order to get accurate values from the titration curve, care should be taken to make many measurements around the equivalence point. One way to accomplish this is to decrease the volume of base you add when the pink color begins to persist in your solution. Once you have passed the equivalence point, the solution will remain pink. The CBL will record volume and pH data for you. From the titration curve, you will determine the equivalence point volume and pH at the volume half-way to the equivalence point. Using this information and the mass of unknown acid used, you will calculate the molar mass and pK_a for your unknown acid. You will determine which acid you were assigned by comparing your data to the table provided at the end of the procedure.

Procedure

1. Work in groups. Wear safety glasses/goggles.

2. Weigh out your unknown acid using weigh paper. Weigh out 0.2 g of your unknown acid. Remember to record the actual mass to two decimal places on your data sheet. **CAUTION:** *Handle the solid acid and its solution with care. Acids can harm your eyes, skin, and respiratory tract.*

3. Transfer the unknown acid to a 100-mL beaker and dissolve it in about 25 mL of distilled water. The volume of water is unimportant, since the number of moles of acid is the only important value for calculations.

4. Add 2 drops of phenolphthalein indicator to your unknown acid solution and place under the hood of your lab bench.

5. Obtain approximately 75 mL of 0.1000M NaOH in a 150-mL beaker and a buret from the cabinet in the back of the lab. **CAUTION:** *Sodium hydroxide solution is caustic. Avoid spilling it on your skin or clothing.*

6. Using your funnel, add about 5 mL of base solution to the buret and rinse it. Discard the base solution through the wide mouth of the buret.

7. Rinse the buret with another 5 mL of base solution. This time drain some of the base solution through the stop-cock leaving some in the buret. Make sure that there are no bubbles left in the tip of the buret. Use a butterfly clamp to hold the buret. See Figure 2.

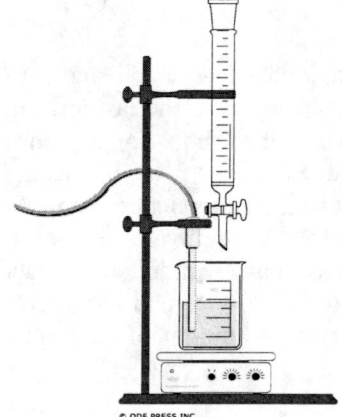

Figure 2

8. Add the remaining base solution to the buret so that the meniscus is above the zero mark. Drain enough base solution out to lower the meniscus to the zero mark. This will allow you to input the exact volume you read from the buret into the CBL. If you start at a volume other than zero mark, you will have to subtract that starting volume from every reading you make before entering it on the CBL.

9. **Calibration of pH probe and CBL.** In order to measure pH, we must first calibrate the CBL pH scale. Obtain pH 4.0 buffer and pH 7.0 buffer.

    o Make sure the CBL is plugged into an outlet and to the TI graphing calculator. Plug the pH electrode into the Channel 1 slot on the CBL. Turn on the calculator and CBL.

    o On the calculator, press PRGM and select CHEMBIO. Press ENTER to go to the main menu.

    o Select SET UP Probes from the main menu. Enter "1" as the number of probes. Select pH from the SELECT PROBE menu.

    o Enter "1" as the channel number and select PERFORM NEW from the CALIBRATION menu.

- Remove the pH electrode from the storage container and rinse it with distilled water. Dry the probe lightly with a paper towel and place it in the pH 4.0 buffer solution.
- Wait 30 seconds for the system to warm-up and press ENTER. When the CBL has stabilized, press TRIGGER on the CBL.
- Type 4.0 on the calculator and press ENTER.
- Remove the pH electrode from the pH 4.0 buffer and rinse and dry it. Place the electrode in the pH 7.0 buffer.
- Once the CBL has stabilized, press TRIGGER on the CBL and type 7.0 in the calculator and press ENTER.
- Press ENTER to complete the calibration process. Remove the pH electrode from the pH 7.0 buffer solution, rinse and dry it, and place it in the unknown acid solution.

10. **Prepare the CBL and collect data.**
    - On the main menu, select COLLECT DATA.
    - Select TRIGGER/PROMPT from the DATA COLLECTION menu and follow the directions on the screen to allow the system to warm-up, then press ENTER.
    - You are now ready to begin the titration. This process will go faster if one person manipulates the buret, one person reads the buret, and another person operates the calculator. Press ENTER after the pH electrode has warmed up. Before adding any base solution, record a pH measurement by allowing the reading to stabilize and pressing TRIGGER. Enter 0.00 as the volume of base added on the calculator and press ENTER.
    - Select MORE DATA and add base solution in 0.5 mL increments, pressing TRIGGER after the reading stabilizes after each addition and entering the cumulative volume of base added (to two decimal places) on the calculator. Make sure you stir the solution well with each addition.
    - When the pink color in the acid solution begins to persist (~10-15 mL of base added), decrease the amount of base added in each increment to 2 drops.
    - When the pink color in the acid solution does not go away with stirring, go back to 0.5 mL increments until pH 11 or 25 mL of base has been reached.
    - Select STOP AND GRAPH from the DATA COLLECTION menu after the last base addition. The graph displayed will show pH (Y-axis) versus volume of base added (X-axis). You can use the trace function on the calculator to move along the titration curve and see specific pH and base volume data points.

11. Drain the buret to the 25.00 mL mark. Repeat the experiment using the remaining NaOH in your buret. You will need to subtract 25.00 mL from each volume reading before you enter it on the calculator. Calculate the molar mass and pK_a for each titration and average your results.

12. Dispose of contents of the beaker and buret as directed by your instructor. Rinse the pH electrode with distilled water and place it in the storage container.

Experiment 17　　　　　　　　　　　Identification of an Unknown Acid By Titration

Worksheet 17

Name _____

Date _____ Lab Instructor/Section _____

Pre-lab	____/20
Data	____/20
Post-lab	____/20
Safety/Part.	____/40
Total	____/100

Data sheet

Molar Mass

1. Unknown assigned _____

2. Mass of unknown acid _____

3. Volume of 0.1000 M NaOH at equivalence point _____

4. Moles of NaOH at equivalence point _____

5. Moles of unknown acid in beaker _____

6. Molar Mass of unknown acid _____

pK_a

1. Volume of NaOH at equivalence point _____

2. Volume of NaOH half-way to equivalence point _____

3. pH at half-way to equivalence point _____

4. pK_a of unknown acid (Table on page 17-8) _____

Identification of unknown acid

Based on the values calculated above, which of the acids from the table on the back of this page is your unknown?

Experiment 17 — Identification of an Unknown Acid By Titration

Acid	Molar Mass	pK$_a$
Potassium hydrogen phthalate (abbreviated KHP)	204 g/mol	5.41
Potassium hydrogen sulfate (KHSO$_4$)	136 g/mol	1.92
Hydroxylamine HCl	69.5 g/mol	5.95
Lactic Acid	90.0 g/mol	3.86

Questions

1. What is the pH of the solution obtained by mixing 35.00 mL of 0.250 M HCl and 35.00 mL of 0.125 M NaOH?

2. What is the pH of a solution prepared by mixing 20.0 mL of 0.5 M NaOH and 35.0 mL of 0.50 benzoic acid solution? (Benzoic acid is monoprotic with a K_a of 6.5 x 10^{-5}.)

Experiment 17 — Identification of an Unknown Acid By Titration

Pre-Laboratory 17

Name _____

Date _____ Lab Instructor/Section _____

Pre-laboratory

Given the followuing information: 1.6 g of an unknown monoprotic acid (HA) required 50.80 mL of a 0.35 M NaOH solution to reach the equivalence point. Calculate the molar mass of the acid. Show all calculations.

The pH was measured to be 3.86 at 25.40 mL. Calculate the K_a of the unknown acid. Show all calculations.

Experiment 17 Identification of an Unknown Acid By Titration

Vol, mL	pH

Vol, mL	pH

Vol, mL	pH

Vol, mL	pH

Vol, mL	pH

Vol, mL	pH

Experiment 18

Identification of Metals by Measuring Potentials of Micro-Voltaic Cells

Objective

The identity of five unknown metals will be determined by measuring the difference in potential (voltage) between each pair of all the possible half-cell reactions.

Equipment

CBL system
TI graphing calculator
AC adapter
TI Voltage probe
Sand paper
Forceps

1 M solutions of M_1^{2+}, M_2^{2+}, ..., and M_5^{2+}
1 X 1 cm metals M_1, M_2, M_3, M_4 and M_5
1 M NaNO$_3$
One piece of filter paper, 11.0 cm diameter
One glass plate, 15 X 15 cm

Safety Precautions

Handle all the solutions with care. Some are poisonous and some cause hard to remove stains. If a spill occurs, ask you teacher how to clean up safely.

Principles

A **voltaic cell** is an apparatus that produces electrical energy directly from the chemical energy released in a reduction-oxidation (redox) reaction. For example, a redox reaction will occur when a piece of metallic magnesium is immersed into a solution of nickel(II) nitrate. Equation 1 shows the spontaneous chemical reaction.

$$Mg(s) + Ni^{2+}(aq) \rightarrow Mg^{2+}(aq) + Ni(s) \qquad (1)$$

This chemical reaction can be separated into two half-reactions, reactions that show either the oxidation (lose of electrons) or reduction (gain of electrons) portion of the redox reaction, including the electrons.

$$Mg(s) \rightarrow Mg^{2+}(aq) + 2e^- \quad \text{(Oxidation)} \tag{2}$$

$$Ni^{2+}(aq) + 2e^- \rightarrow Ni(s) \quad \text{(Reduction)} \tag{3}$$

This redox reaction can be converted into a voltaic cell by physically separating the reactants and directing the electron transfer through an electrical conductor. The resulting electric current can then be used to perform work. For a more detailed description of a voltaic cell refer to the chapter in your text dealing with electrochemistry.

The electrical driving force that pushes the electrons generated in the oxidation half-cell toward the electrode where reduction takes place is termed the **electromotive force (emf)**. The greater the potential difference between the electrons at the two electrodes, the larger the emf. The **SI** unit for emf is the **volt (V)**. Therefore, the emf of a cell is measured with a voltmeter. The **potential** of the cell (E_{cell}) is the potential energy difference between the electrodes of a voltaic cell. Since the cell potential is measured in units of volts, cell voltage is often used interchangeably with cell potential or emf. The **standard potential** of a cell ($E°_{cell}$) is the voltage that is measured when all of the reactants and the products in the redox reaction are in their standard states - i.e., solids, liquids, and gases in the pure state at 1 atm pressure, and solutes at a concentration of 1.0 M. The measured standard cell potential for Equation 1 is +2.12V.

$$Mg(s) + Ni^{2+}(1M) \rightarrow Mg^{2+}(1M) + Ni(s)$$
$$E°_{cell} = +2.12 \text{ V} \tag{4}$$

The voltage of the cell is positive indicating that the cell reaction occurs spontaneously as written. This spontaneous flow of electrons always occurs from the negative electrode through the external circuit to the positive electrode. For the magnesium/nickel cell, shown above the positive emf means that the electrons flow spontaneously from the magnesium electrode to the nickel electrode. In this example the nickel electrode is at a positive voltage with respect to the magnesium electrode.

It is important to realize that experimental measurements record only differences in the potentials of two half-reactions and that absolute potentials cannot be measured. If you were to replace the magnesium half-cell with the hydrogen electrode the reaction for a spontaneous reaction and measured potential is:

$$2H^+(1M) + Ni(s) \rightarrow H_{2(g,\ 1\ atm)} + Ni^{2+}(1M)$$
$$E°_{cell} = +0.25 \text{ V} \tag{5}$$

Notice that in this case the hydrogen electrode is at a positive voltage with respect to the nickel electrode. Measurement of the potential of a magnesium half-cell with the standard hydrogen half-cell yields a much more positive value.

$$Mg(s) + 2H^+(1M) \rightarrow Mg^{2+}(1M) + H_{2\ (g,\ 1\ atm)}$$
$$E°_{cell} = +2.37 \text{ V} \tag{6}$$

It is important to notice that Equation 6 is the sum of Equations 4 and 5, and its potential is the sum of the potentials of the other two cells. Experimental measurements of many cell potentials have confirmed that **cell potentials are additive**.

Because cell potentials are additive, one can calculate the potential of a cell using the measured potentials of other cells. Furthermore, it is possible to assign standard potentials to half-reactions by arbitrarily defining the standard potential of one particular half-reaction. Scientist have chosen the reduction of hydrogen ions to hydrogen gas as a reference and assigned it a standard potential of 0.00 V.

$2H^+(1M) + 2e^- \rightarrow H_{2\ (g,\ 1\ atm)}$
$$E°_{cell} = 0.00\ V \qquad (7)$$

By arbitrarily assigning the hydrogen electrode to 0.0 V one can construct a cell that combines the standard hydrogen half-cell and any other standard half-cell and measure its voltage. Since we have chosen zero as the potential of the standard hydrogen half-cell, the observed voltage of the cell is attributed to the reaction in the second half-cell. A table of standard reduction potentials for a large number of half-reactions, all referenced to the reduction of hydrogen ions to hydrogen gas, can be found in most general chemistry textbooks. The table is arranged so that all of the half reactions are written as reductions.

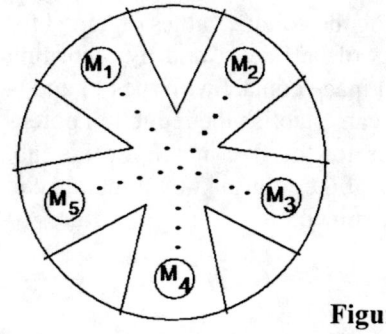

Figure 1

In this lab you will establish a table of reduction potentials for copper (M_1) and four unknown metals using micro-voltaic cells. Half-cells are normally produced by placing a piece of metal into a solution containing a cation of the metal (e.g., Cu metal in a solution of $CuSO_4$). In this micro-version of a voltaic cell, the half cell will be a small piece of metal placed into 3 drops of solution on a piece of filter paper. The solution contains the cation of the solid metal. Figure 1 shows the arrangement of half-cells on the piece of filter paper. A porous barrier or a salt bridge normally separates the two half-reactions. Here, the salt bridge will be several drops of aqueous $NaNO_3$ placed on the filter paper between the two half-cells. Using the CBL as a voltmeter, the (+) lead makes contact with one metal and the (-) lead with another. If a positive voltage is recorded on the screen, you have connected the cell correctly. The metal attached to the (+) lead is the cathode (reduction) and thus has a higher, more positive, reduction potential. The metal attached to the (-) lead is the anode (oxidation) and has a lower, more negative, reduction potential. If you get a negative voltage reading, then you must reverse the leads.

By comparing the voltage values obtained for several pairs of half-cells, and by recording which metal made contact with the (+) and (−) leads, you can establish the reduction potential sequence for the five metals in this lab. The identity of the four unknown metals can then be determined.

Procedure

1. Obtain and wear goggles.

2. Prepare the voltage probe for data collection.

    - Plug the voltage probe directly into Channel 1 of the CBL.
    - Use the link cable to connect the CBL System to the TI Graphing Calculator. Firmly press in the cable ends.

3. Turn on the CBL unit and the calculator. Start the CHEMBIO program and proceed to the MAIN MENU.

4. Set up the calculator and CBL for the voltage probe and a calibration (in volts).

    - Select SET UP PROBES from the MAIN MENU.
    - Enter "1" as the number of probes.
    - Select VOLTAGE from the SELECT PROBE menu.
    - Enter "1" as the channel number.

5. Set up the calculator and CBL for data collection.

    - Select COLLECT DATA from the MAIN MENU.
    - Select MONITOR INPUT from the DATA COLLECTION menu.
    - The voltage reading is displayed on the screens of the CBL and the TI calculator. No readings are stored when using the MONITOR INPUT mode.

6. Obtain a piece of filter paper and draw five small circles with connecting lines, as shown in Figure 1. Using a pair of scissors, cut wedges between the circles as shown. Label the circles M_1, M_2, M_3, M_4, and M_5. Place the filter paper on top of the glass plate.

7. Obtain 5 pieces of metal, M_1, M_2, M_3, M_4, and M_5. Sand each piece of metal on both sides. Place each metal near the circle with the same number.

8. Place 3 drops of each solution on its circle (M_1^{2+} on M_1, etc.). Then place the piece of metal on the wet spot with its respective cation. The top side of the metal should be kept dry. Then add several drops of 1 M $NaNO_3$ to the line drawn between each circle and the center of the filter paper. Be sure there is a continuous trail of $NaNO_3$ between each circle and the center. You may have to periodically dampen the filter paper with $NaNO_3$ during the experiment. **CAUTION:** *Handle these solutions with care. Some are poisonous and some cause hard-to-remove stains. If a spill occurs, ask your teacher how to clean up safely.*

9. Use metal M_1 (copper) as the reference metal. Determine the potential of four cells by connecting M_1 to M_2, M_1 to M_3, M_1 to M_4, and M_1 to M_5. This measurement is made by bringing the (+) lead in contact with one metal and the (−) lead in contact with the other. If the displayed voltage on the CBL or calculator screen is (−), then reverse the leads. Wait about 5 seconds to take a voltage reading, and record the (+) value appearing on the CBL or calculator screen in Data Table 1 (round to the nearest 0.01 V). Also record which metal is the (+) terminal and which is (−), when the voltage value is positive. Use the same procedure and measure the potential of the other three cells, continuing to use M_1 as the reference electrode.

Experiment 18 — Identification of Metals by Measuring Micro-Voltaic Cells

10. Go to Step 1 of Processing the Data. Use the method described in Step 1 to rank the five metals from the lowest (–) reduction potential to the highest (+) reduction potential. Then *predict* the potentials for the remaining six cell combinations.

11. Now return to your work station and *measure* the potential of the six remaining half-cell combinations using the CBL and TI calculator. If the NaNO3 salt bridge solution has dried, you may have to re-moisten it. Record each measured potential in Data Table 3.

12. When you have finished, use forceps to remove each of the pieces of metal from the filter paper. Rinse each piece of metal with tap water. Dry it and return it to the correct container. Remove the filter paper from the glass plate using the forceps, and discard it as directed by your teacher. Rinse the glass plate with tap water, making sure that your hands do not come in contact with wet spots on the glass. Press [+] to quit MONITOR INPUT.

Processing the Data

1. After finishing Step 9 in the procedure, arrange the five metals (including M_1) in Data Table 2 from the lowest reduction potential at the top (most negative) to the highest reduction potential at the bottom (most positive). Metal M_1, the standard reference, will be given an arbitrary value of 0.00 V. If the other metal was correctly connected to the *negative* terminal, it will be placed *above* M_1 in the chart (with a negative E° value). If it was connected to the positive terminal, it will be placed below M_1 in the chart (with a positive E° value). The numerical value of the potential relative to M_1 will simply be the value that you measured on the CBL. Record your results in Data Table 2.

    Then calculate the *predicted* potential of each of the remaining cell combinations shown in Data Table 3, using the reduction potentials you just determined (in Data Table 2). Record the predicted cell potentials in Data Table 3. Return to Step 11 in the procedure and finish the experiment.

2. Calculate the % error for each of the potentials you measured in Step 11 of the procedure. Do this by comparing the measured cell potentials with the predicted cell potentials in Data Table 3.

3. You can determine the identity of metals M_2 through M_5 using a reduction potential chart in your textbook or the shorter one at the end of this experiment. Remember that hydrogen, H_2, has a reduction potential of 0.00 V on these charts, but that this experiment uses the copper half reaction as the standard. Locate copper, M_1, on the chart, and then determine the likely identity of each of the other metals using your experimental reduction potential sequence in Data Table 2. Note: One of the unknown metals has a +1 oxidation state; the remainder of the metals have +2 oxidation states.

Experiment 18　　　　　Identification of Metals by Measuring Micro-Voltaic Cells

Worksheet 18

Name _____

Date _____　Lab Instructor/Section _____

Pre-lab	____/20
Data	____/20
Post-lab	____/20
Safety/Part.	____/40
Total	____/100

DATA TABLE 1

Voltaic Cell (metals used)	Measured Potential (V)	Metal Number of (+) Lead	Metal Number of (−) Lead
M_1 / M_2			
M_1 / M_3			
M_1 / M_4			
M_1 / M_5			

DATA TABLE 2

Metal (M_x)	Lowest (−) Reduction Potential, E_i (V)
	Highest (+) Reduction Potential, E_i (V)

- 185 -

Experiment 18   Identification of Metals by Measuring Micro-Voltaic Cells

DATA TABLE 3

	Predicted Potential (V)	Measured Potential (V)	Percent Error (%)
M_2 / M_3			
M_2 / M_4			
M_2 / M_5			
M_3 / M_4			
M_3 / M_5			
M_4 / M_5			

Metal Identity

M1   Copper_____

M2   _____

M3   _____

M4   _____

M5   _____

Questions

1. Define the following: salt bridge, emf, cell potential, and voltaic cell.

2. Write a redox equation using nickel and magnesium that gives a cell potential of −2.12 V at standard conditions.

Experiment 18   Identification of Metals by Measuring Micro-Voltaic Cells

Worksheet 18

Name _____

Date _____ Lab Instructor/Section _____

Pre-laboratory Exercise

1. Write the chemical equations that occur in the following cells:

    Pb | Pb(NO$_3$)$_2$ || AgNO$_3$ | Ag

    Zn | ZnCl$_2$ || Pb(NO$_3$)$_2$ | Pb

    Pb | Pb(NO$_3$)$_2$ || NiCl$_2$ | Ni

2. Using the reduction potential chart at the end of this experiment. Determine the value of the following half reactions if the standard was referenced to the reduction of I$_2$ solid to I$^-$ ions (i.e. I$_2$(s) + 2e$^-$ → 2I$^-$(aq)   E° = 0.00 V):

    Cl$_2$(g) + 2e$^-$ → 2Cl$^-$(aq)   E° = _____

    Pb^{2+}(aq) + 2e$^-$ → Pb(s)   E° = _____

    2H$^+$(aq) + 2e$^-$ → H$_2$(g)   E° = _____

Reduction Potential Table at 25 °C

Reduction half-reactions		E° (V)
$F_2(g) + 2e^-$	$\rightarrow 2F^-(aq)$	2.87
$Ce^{4+}(aq) + e^-$	$\rightarrow Ce^{3+}(aq)$	1.61
$Cl_2(g) + 2e^-$	$\rightarrow 2Cl^-(aq)$	1.36
$Br_2(l) + 2e^-$	$\rightarrow 2Br^-(aq)$	1.06
$Ag^+(aq) + e^-$	$\rightarrow Ag(s)$	0.80
$Fe^{3+}(aq) + e^-$	$\rightarrow Fe^{2+}(aq)$	0.77
$I_2(s) + 2e^-$	$\rightarrow 2I^-(aq)$	0.54
$Cu^{2+}(aq) + 2e^-$	$\rightarrow Cu(s)$	0.34
$Sn^{4+}(aq) + 2e^-$	$\rightarrow Sn^{2+}(aq)$	0.15
$2H^+(aq) + 2e^-$	$\rightarrow H_2(g)$	0.000
$Pb^{2+}(aq) + 2e^-$	$\rightarrow Pb(s)$	-0.126
$Ni^{2+}(aq) + 2e^-$	$\rightarrow Ni(s)$	-0.25
$Cr^{3+}(aq) + e^-$	$\rightarrow Cr^{2+}(aq)$	-0.41
$Fe^{2+}(aq) + 2e^-$	$\rightarrow Fe(s)$	-0.44
$Zn^{2+}(aq) + 2e^-$	$\rightarrow Zn(s)$	-0.76
$Al^{3+}(aq) + 3e^-$	$\rightarrow Al(s)$	-1.66
$Mg^{2+}(aq) + 2e^-$	$\rightarrow Mg(s)$	-2.37

Experiment 19

The Solvay Process-Preparation of Sodium Bicarbonate

Objective

To prepare $NaHCO_3$ from $NaCl$, CO_2 and NH_3.

Equipment and Chemicals

NaCl
Acetone
Ammonium hydroxide
Dry ice

Buchner funnel
5.5 cm filter paper
Large cork
Erlenmeyer flask

Safety Precautions

Wear approved eye protection. Avoid spilling chemicals. Handle the ammonia solution carefully and under the hood. Handle the dry ice only briefly as it is very cold and can "burn" your skin.

Principles

The commercial production of sodium bicarbonate is accomplished by the Solvay process. The chemical reactions involved in the production of sodium bicarbonate are:

$$CaCO_3(s) \rightarrow CaO(s) + CO_2(g) \text{ (at elevated temperature)}$$

$$CO_2(g) + NH_3(aq) + H_2O \rightarrow NH_4^+(aq) + HCO_3^-(aq)$$

$$Na^+(aq) + HCO_3^-(aq) \rightarrow NaHCO_3(s)$$

where (s), (g), (aq) mean. solid, gas and. aqueous (in water solution), respectively. For an industrial process to be successful, the reactions must all go to form exclusively products (go to completion), an equilibrium situation, especially one that is mainly to the left, is negative. These three reactions illustrate some of the important forces that drive chemical reactions to completion. If a product of a reaction is removed as it is formed this tends to drive the reaction to completion. In the first reaction, this removal is accomplished by the formation of gas, and in the third reaction the formation of solid removes the bicarbonate ion from the reaction solution. Acid with a base reaction generally proceed largely toward completion. In the second reaction above we have an acid-base reaction; the CO_2 reacts with water to form carbonic acid (H_2CO_3) that is deprotonated by the base ammonia.

In the actual process, the solution used to absorb the carbon dioxide is saturated in both NH_3 and NaCl (which serves as a source of Na^+). This means that the final solution after separation of the $NaHCO_3$ is an aqueous solution of NH_4Cl.

Since cost is a major factor in a commercial processes, the by-products of these reactions are used in the formation of the starting solution. The CaO is first hydrated producing slaked lime

$$CaO(s) + H_2O \rightarrow Ca(OH)_2(s)$$

Addition of this compound to the ammonium chloride solution produces a solution of $CaCl_2$ and $NH_3(g)$

$$Ca(OH)_2 + 2\,NH_4Cl \rightarrow$$
$$CaCl_2 + 2\,NH_3(g) + H_2O$$

The ammonia gas is dissolved in a fresh sodium chloride solution, and used in the next batch. By recycling the ammonia in this manner, the cost of the process is significantly reduced, and a single by-product, $CaCl_2$ is formed.

In this experiment sodium bicarbonate is prepared by the Solvay process. However, for convenience, the source of carbon dioxide used will be dry ice (solid CO_2)

Procedure

On weighing paper weigh approximately 15 g of sodium chloride to 0.1 g accuracy. Record the weight and transfer it to an Erlenmeyer flask. Add 50 mL of concentrated ammonia solution to the flask and stopper loosely with a rubber stopper.
CAUTION: Handle the ammonia solution carefully and under the hood. The ammonia solution is corrosive with very irritating vapors. Swirl the flask to speed the dissolving of the sodium chloride. If necessary, add more ammonia solution in 5 mL increments until all of the salt has dissolved.

Obtain a 150 mL beaker of "dry ice" pellets (handle the dry ice carefully, because it is very cold and it can burn your hands) and add it to the contents of the flask in small increments, swirling the flask under the hood. DO NOT HAVE THE FLASK STOPPERED during this addition since large volumes of gas are produced. Note: although the temperature of solid CO_2 is $-78°C$ the contents of the flask become warm from the exothermic reaction occurring. Stop the addition of the "dry ice" when the contents begin to cool and a white solid forms. When all the solid dry ice added to the reaction mixture has sublimed away, filter the solid by suction using a Buchner funnel, wash the solid with two 20 mL portions of acetone (use the correct procedure described in the aspirin synthesis) and draw air through the solid to dry.

Transfer the dried solid to a preweighed paper, and determine its mass. Calculate the percent yield obtained, based on the weight of NaCl used.

Test your product by adding a small amount of it to some dilute hydrochloric acid solution. Report your observations and write an equation for the reaction that occurs.

Follow the directions of your instructor for the disposition of the remainder of your product.

Experiment 19 The Solvay Process: Preparation of Sodium Bicarbonate

Worksheet 19

Name _____

Date _____ Lab Instructor/Section _____

Data sheet

Pre-lab	____	/20
Data	____	/20
Post-lab	20	/20
Safety/Part.	____	/40
Total	____	/100

Mass of NaCl and paper _____ g

Mass of paper _____ g

Mass of NaCl _____ g

Mass of product and paper _____ g

Mass of paper _____ g

Mass of product _____ g

Theoretical yield (based on NaCl) _____ g

Percent yield _____ %

Calculations:

Observation when observing the reaction of your product and HCl

Write the equation for the reaction of your product and HCl.

Experiment 19 — The Solvay Process: Preparation of Sodium Bicarbonate

Pre-Laboratory 19

Name _____

Date _____ Lab Instructor/Section _____

Pre-laboratory

1. In a preparation of $NaHCO_3$, 15.2 g of NaCl is used along with excess NH_3 and excess CO_2. The student isolates 20.2 g of $NaHCO_3$. Calculate the theoretical yield and the percent yield.

Experiment 20

Determining the Rate Law for the Reaction of Crystal Violet with OH⁻

Objective

The objective of this experiment is to introduce the student to techniques of determining individual reaction order by graphing data according to different integrated rate law formats. A second objective is for students to observe the slow decomposition visually thereby gaining first hand evidence of reaction rate.

Equipment and chemicals

CBL2 system with calculator
Plastic cuvettes
50-mL beaker
0.10 M NaOH

Colorimeter
Two 10-mL graduated cylinders
Stirring rod
2.0×10^{-5} M crystal violet

Safety Precautions

Wear approved eye protection.

Crystal violet solutions may cause skin and eye irritation. Sodium hydroxide solutions are caustic and will cause skin burns and are extremely hazardous to your eyes. Any skin contact with either chemical should be immediately washed. Wash your hands with soap and water before leaving the lab.

Principles

In this experiment, you will observe the reaction between crystal violet and sodium hydroxide. Crystal violet is available as the chloride salt, which completely dissociates in water. The equation for the reaction with hydroxide is:

A simplified version of the equation is:

$$CV^+ + OH^- \longrightarrow CVOH$$

crystal violet cation (violet)   hydroxide ion (colorless)   (colorless)

The rate law for this reaction is in the form:

$$\text{rate} = k[CV^+]^x [OH^-]^y,$$

where k is the rate constant for the reaction, x is the order with respect to crystal violet (CV^+), and y is the order with respect to the hydroxide ion. To simplify the initial data treatment, the hydroxide ion concentration is 4000 times as large as the concentration of crystal violet. During the course of the reaction, the hydroxide-ion concentration remains essentially constant so the experimental rate of reaction depends only on the crystal violet concentration.

In this experiment, you will acquire data to find the order with respect to crystal violet (x), but not the order with respect to hydroxide (y).

As the reaction proceeds, a violet-colored reactant (crystal violet) reacts with hydroxide ion to form a colorless product. Using the green (565 nm) light source of a colorimeter, you will monitor the absorbance of the crystal violet solution with time. We will assume that absorbance is proportional to the concentration of crystal violet (Beer's law). With the concentration data the following three graphs can be plotted:

• Concentration vs. time: A linear plot indicates a zero-order reaction (k = –slope).
• ln Concentration vs. time: A linear plot indicates a first order reaction (k = –slope).
• 1/Concentration vs. time: A linear plot indicates a second order reaction (k = slope).

Once the order with respect to crystal violet has been determined, you will also be finding the rate constant, k, from the experimental data.

Procedure

Integrated Rate Laws: Determination of the order in Crystal Violet.

1. Plug the colorimeter into Channel 1 of the CBL2 System.

2. Turn on the calculator. Press **[PRGM]** and select CHEMBIO. Press **[ENTER]**. "prgmCHEMBIO" will appear. Press **[ENTER]**. "Vernier Software Biology and Chemistry with the CBL" will appear. Press **[ENTER]** again to go to the MAIN MENU.

3. Set up the calculator and CBL for the colorimeter.

• Select: **SET UP PROBES** from the MAIN MENU. Press **[ENTER]**.
• Enter "1" as the number of probes. Press **[ENTER]**.
• Select: **COLORIMETER** from the SELECT PROBE menu. Press **[ENTER]**.
• Enter "1" as the channel number. Press **[ENTER]**.

4. The calculator will direct you to press the arrows on the colorimeter to select the appropriate wavelength. Select 565 nm. Press [ENTER] on the calculator. This takes you to the calibration process.

5. Place a cuvette filled with deionized water in the CBL. Press **[ENTER]**. Press the "CAL" button on the

colorimeter. Once the red LED on the colorimeter stops flashing, press **[ENTER]** on the calculator. You should return to the MAIN MENU.

6. Set up the calculator and CBL for data collection.

   - Select: **COLLECT DATA** from the MAIN MENU. Press **[ENTER]**.
   - Select: **TIME GRAPH** from the DATA COLLECTION menu. Press **[ENTER]**.
   - Enter 8 for time between samples and press **[ENTER]**.
   - Enter 55 as the number of samples and Press **[ENTER]**.
   - The experiment length should be 440.00 seconds. Press **[ENTER]**.
   - Select: **USE TIME SETUP** from the MAIN MENU. Press **[ENTER]**.
   - For Ymin enter 0 and Press **[ENTER]**. For Ymax enter 1 and Press **[ENTER]**.
   - For Yscl enter 0.2 and Press **[ENTER]**. The system is now ready to collect data. **Do not** press **[ENTER]** until you are ready to begin the experiment.

7. Use a 10-mL graduated cylinder to measure out 10.0 mL of 0.10 M NaOH solution. **Caution**: Sodium hydroxide solution is caustic. Avoid spilling it on your skin or clothing.

8. Use another 10-mL graduated cylinder to measure out 10.0 mL of 2.0×10^{-5} M crystal violet solution. **Caution**: Crystal violet is a biological stain. Avoid spilling it on your skin or clothing.

9. You are now ready to begin monitoring data. To initiate the reaction, simultaneously pour the 10-mL portions of crystal violet and sodium hydroxide into a 50-mL beaker and stir the reaction mixture with a stirring rod. Rinse the cuvet with ~1-mL of the reaction mixture and then fill it ¾ full. Place the cuvet in the cuvet slot of the colorimeter with the clear side facing the path of the light and close the lid. Wait for about 10 seconds, then press **[ENTER]** to begin collecting data. During the ~7-minute data collection, observe the solution in the beaker as it continues to react. Data collection stops after ~7 minutes. Remove the cuvet from the colorimeter compartment, and discard the contents of the beaker and cell as directed by your instructor. Press **[ENTER]**. Your data will appear in a graph. Press **[ENTER]** to leave the graph. Select **[NO]** and press **[ENTER]** to return to the MAIN MENU.

10. Press **[7]** to exit the program and work with the data in the lists in the calculator. The time data will be in the L1 list and the absorbance (concentration) data will be in L2. To access this data, press **[STAT]** and **[ENTER]**. In L3, enter LN(L2) and in L4 enter 1/L2. We now have data for zero, first and second order integrated rate laws.

Data Analysis

1. Plot the raw concentration data (L2) versus time (L1) using the **[STAT PLOT]** function on the calculator. Make sure all of the Plots are OFF, except for Plot1. Select Plot1 and press **[ENTER]** to modify the settings. The Xlist should be L1 and the Ylist should be L2. Press **[ZOOM]** and select **ZoomStat**. Press **[ENTER]**. Your graph should appear. Sketch a picture of your graph on your data worksheet. Repeat the graphing process changing the Ylist to L3 and then to L4 to get each of the graphs below.

   - Zero Order: If the graph of concentration vs. time is linear, the reaction is zero order.

- **First Order:** If the graph of ln concentration versus time gives a straight line, then the reaction is first order.
- **Second Order:** If the graph of 1/concentration versus time gives a straight line, then the reaction is second order.

2. To better determine which order best fits the data, we can calculate r^2 values, which tell us the fraction of variability in the data that is explained by the linear relationship. A value close to 1 indicates a better straight line. We also need the slope and intercept values for the straight line graph and we can get the calculator to do this for us. Press **[STAT]** and scroll to TESTS. Scroll down to LinRegTTest and press **[ENTER]**.

3. Select L1 for Xlist and L2 for Ylist. Scroll down to Calculate and press **[ENTER]**. A list of calculated values will be displayed. We need the slope which is variable b and the r^2. Repeat this changing Ylist to L3 and L4. You should have three slopes, three intercepts, and three r^2 values when you are done. Fill in the table on your worksheet.

4. Determine the order with respect to crystal violet by identifying the integrated rate law associated with the dataset that produced the best straight line.

5. Use the slope of the best fit line to calculate the value of the experimental rate constant.

Experiment 20 Determination...Initial Rate Method

Worksheet 20

Name _____

Date _____ Lab Instructor/Section _____

Pre-lab	____/20
Data	____/20
Post-lab	____/20
Safety/Part.	____/40
Total	____/100

Data sheet

Sketches of Graphs

    A. Concentration (Absorbance) versus time

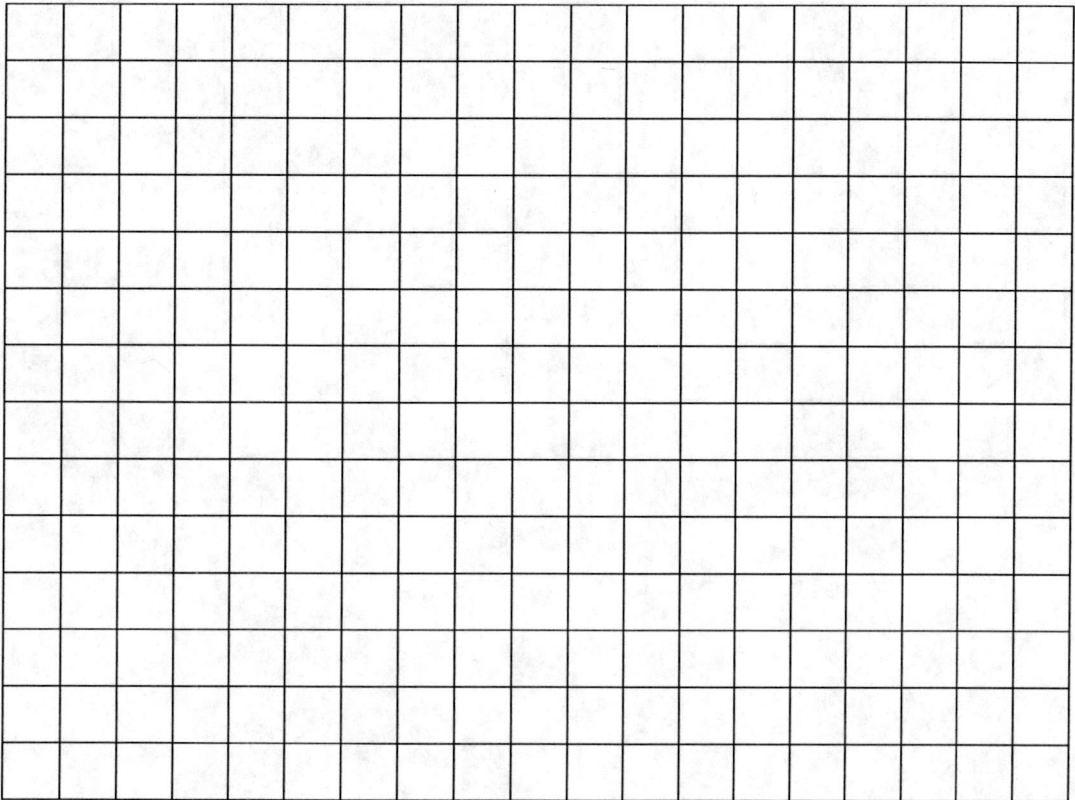

- 199 -

B. ln(Concentration) versus time

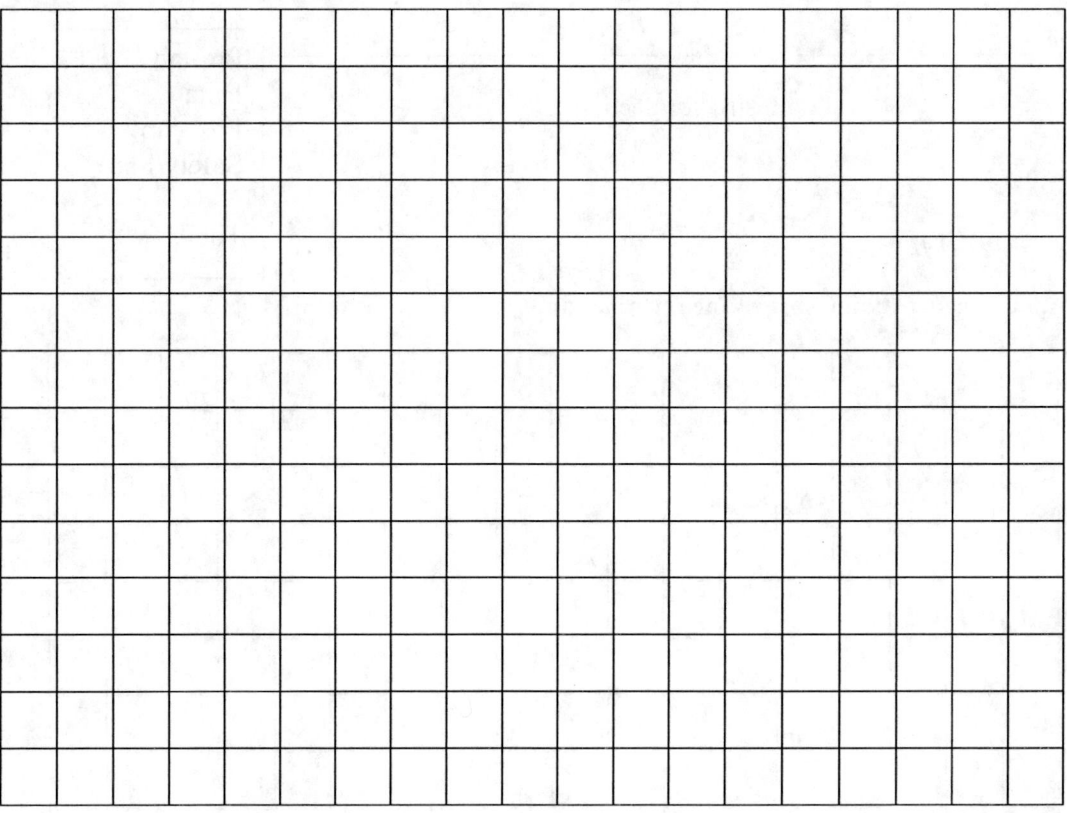

C. 1/concentration versus time

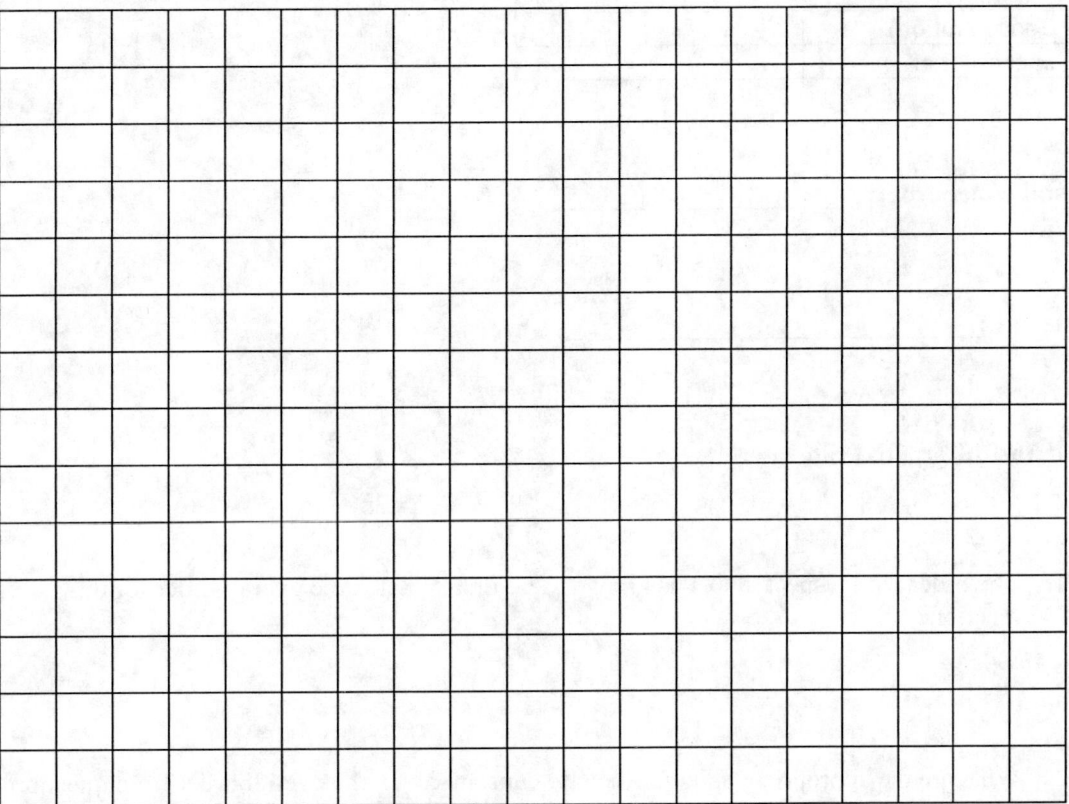

Data Type	Slope	r^2
Concentration		
ln (Concentration)		
1/Concentration		

Crystal violet order: _____

Value of k: _____

Write the differential rate law:

Questions

1. The order with respect to OH^- is known to be first. What is the overall order for this reaction?

2. What are the appropriate units for the rate constant given the overall order from question 1?

3. What effect will doubling the initial concentration of crystal violet have on the rate?

Experiment 20   Determination...Initial Rate Method

Laboratory 20

Name _____

Date _____ Lab Instructor/Section _____

Pre-laboratory

Part A

1. From the following graphs, what is the order of the reactant A? Explain.

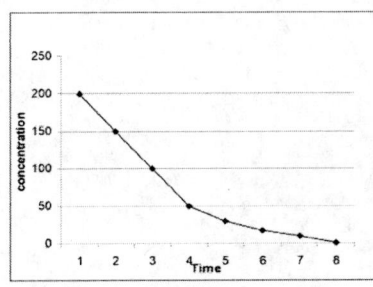 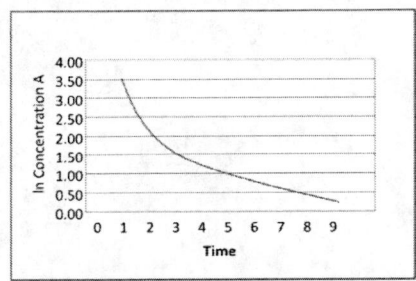 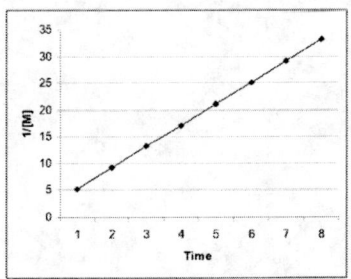

2. Find the value of k from the above graph – **include correct units**.

Additional questions on back of page.
Part B

Exp	[A]	[B]	Rate (M/s)
1	.0319	.0520	4.1×10^{-3}
2	.0319	.104	4.1×10^{-3}
3	.0639	.104	1.63×10^{-2}

- 203 -

1. From the above data, find the order of the reaction with respect to A and the order with respect to B:

2. What is the value of k from the above data? Include correct units.

Part C

1. Given the information in parts A and B, could the two reactions discussed be identical? Explain your answer.